SUSTAINABILITY IN THE SUBURBS

A Green Living Guide for Modern Families

LAURA TROTTA

Published in Australia

Printed in Australia

First Edition

National Library of Australia Cataloguing-in-Publication entry available for this title at nla.gov.au

ISBN: 978-0-6456354-0-9

Cover Design: Ellissa Nagle

Interior Design: Kelly Exeter

Author Photo: Kate Dyer

I dedicate this book to my sons Matthew and Christopher, and to future generations who will inhabit the earth they borrow from their children.

I acknowledge the traditional Country of the Kaurna people of the Adelaide Plains and pay my respect to Elders past, present and future.

I recognise and respect their cultural heritage, beliefs and relationship with the land. I acknowledge that they are of continuing importance to the Kaurna people living today.

I also extend my respect to other Aboriginal and Torres Strait Islander Groups and other First Nations peoples.

I tampendi, ngadlu Kaurna yertangga banbabanbalyarnendi (inbarendi). Kaurna meyunna yaitya mattanya Womma Tarndanyako.

Parnako yailtya, parnuko tappa purruna, parnuko yerta ngadlu tampendi. Yellaka Kaurna meyunna itto yailtya, tappa purruna, yerta kuma burro martendi, burro warriappendi, burro tangka martulyaiendi.

Kumarta yaitya miyurna iyangka yalaka ngadlu tampinthi.

CONTENTS

"What you do makes a difference, and you have to decide what kind of difference you want to make."

—DR JANE GOODALL

INTRODUCTION

"I probably would have preferred to live in my grandmother's day because I wouldn't be afraid of the ozone layer and of the world exploding on me. The world was a much safer place then."
—ME, AGED 11 YEARS, 1989

Much has changed in the three decades since I wrote the above conclusion for a family history school assignment.

Since 1989, global atmospheric carbon dioxide concentrations have increased from approximately 350 parts per million (ppm) to 420 ppm (2022). This is a significant rise from the stable 280 ppm we experienced prior to the industrial revolution.[1]

Global mean atmospheric temperatures in 1989 were 0.27 degrees Celsius (°C) (32.5 degrees Fahrenheit (°F)) above

NASA's baseline 1951-1980 mean.[2] Average global temperatures on Earth are now at least 1.1°C (33.9°F) above that mean.[3]

Our changing climate and ongoing deforestation is resulting in biodiversity being lost at a rate not seen since Earth's last mass extinction. Yet, even though we know that the burning of fossil fuels is a direct cause of climate change, Australian coal production and exports have grown approximately 300%[4] since 1989 and 12 new coal-fired power stations have been commissioned.[5]

Those of us who aren't aware of the numbers above can't escape the daily reports of coral reefs dying, bulldozers razing rainforests, species pushed to the brink of survival, glaciers and polar ice caps melting, cyclones destroying towns and cities, storm surges inundating our coastlines, extreme droughts followed by record-breaking floods and wildfires ravaging our forests and suffocating major cities with toxic smoke.

It's easy to feel overwhelmed and defeated about the poor state of our environment. The problems seem too big, too hard to solve and possibly even too late to solve. So, it's no surprise that many people, including our politicians, (perhaps especially our politicians), bury their heads in the sand and hope that the issues will simply disappear.

The bad news is, the climate crisis and state of our global environment will not fix themselves.

The good news is, you don't need to be an environmental professional, activist or politician to make real, positive change to our planet.

Your actions matter.

You matter.

What our earth desperately needs is a **large mass of people making many seemingly small changes to their lifestyle** and applying these changes consistently with intent and passion. **Positive change on a grand scale will then naturally happen.**

"That's great to know", I hear you say. "But where am I going to start and what can I do?"

In this book I'll guide you through the changes you need to make to reduce your environmental footprint on our earth. The best thing about these simple changes is they won't just improve our environment, they'll improve your health and save you money too.

How we got here and why we need to change

Before I delve into the specifics of what you can do in your own home to help our environment, it's important to have some insight into the current state of our planet and understand the primary drivers of what got us here.

Now, it's not an easy task to rank the top environmental issues facing our planet today. I suspect if every environmental scientist was asked, he or she would come up with a slightly different list. However, the list below is my list, based on my experience of 27+ years studying and working in the fields of environmental science, engineering and sustainability.

*Without a doubt the
biggest threat to our
planet and indeed
our own existence, is
ourselves.*

1. Overpopulation
2. Climate change
3. Habitat and biodiversity loss
4. Water security
5. Ocean system collapse
6. Pollution
7. Waste

Let's take a look at each of these in more detail now.

1. Overpopulation

Without a doubt the biggest threat to our planet and indeed our own existence, is ourselves. Overpopulation is a very real issue as the more people there are on our planet, the more raw materials are consumed to sustain our lives and standard of living.

The sheer number of humans on Earth places stress on every aspect of our environment as we compete for resources with each other and wildlife. Indeed, all other major environmental issues stem from overpopulation: climate change, habitat and biodiversity loss, water scarcity, ocean system collapse, pollution and waste.

In 1900, the world human population stood at 1.6 billion.

As a result of medical advancements, increases in agricultural productivity and availability of cheap energy, the world's human population has grown exponentially since then.

In 2018, a team of scientists concluded that Earth can sustain, at most, 7 billion people at subsistence levels of consumption.[6] Our population passed 7 billion people in October 2011 and is on track to reach 8 billion people in November 2022. Projections by the United Nations suggest that the size of the global human population could grow to almost 11 billion by around 2100.[7]

2. Climate Change

The term 'climate change' refers to a variation in global climate patterns over a long period of time.

The change in our global climate from the 20th century onwards has come primarily from three sources:

1. land use changes (like cutting down forests to create farmland)

2. the release of carbon dioxide and other greenhouse gases (like methane and nitrous oxide) from burning fossil fuels (coal, oil and natural gas) to generate electricity

3. the release of greenhouse gases from manufacturing, mining, transportation, food production and powering buildings.

When released into the Earth's atmosphere, greenhouse gases absorb infrared radiation from the sun and act like an insulating blanket, trapping heat and keeping the Earth warmer than it would be if they were not present. Once released, many of these gases remain in the atmosphere and our climate system for hundreds to thousands of years.

According to an ongoing temperature analysis by scientists at NASA, the average global temperature on Earth has increased by at least 1.1°C (33.98°F) since records began in 1880.[8] Global warming does not mean that temperatures rise everywhere at the same time by the same rate. Australia's climate has warmed in both surface air and surrounding sea surface temperatures by around 1.44°C (34.59°F) since 1910.[9]

Rising temperatures have been recorded on all continents and oceans across the globe. World Meteorological Organization records show that the decade of 2011-2020 was the world's warmest decade on record. Since the 1980s, each decade has been warmer than the previous one and this trend is expected to continue.[10]

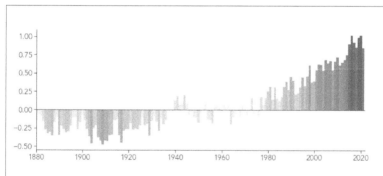

Figure 1: Global Temperature Anomaly (°C compared to the 1951 – 1980 average) (Source: NASA Earth Observatory)

*We can avoid the
worst effects of climate
change if carbon dioxide
emissions are reduced to
an acceptable level.*

The consequences of climate change are profound and include:

» increases in the frequency and intensity of extreme weather events such as droughts, bushfires, floods and storms

» melting of ice sheets which lead to rising sea levels that pose a significant risk to coastal communities

» the world's oceans becoming too acidic to support coral reefs and other calcifying marine organisms

» loss of habitat

» catastrophic ecological collapse

» biodiversity loss

» food system collapse

» mass migration of people (climate change refugees).

Many climate scientists believe it may be too late to undo the damage already inflicted on the environment by climate change. But they agree we can avoid the worst effects of climate change if carbon dioxide emissions are reduced to an acceptable level.

The Paris Agreement (2015) calls for all countries to limit global warming to well below 2°C (35.6°F), preferably 1.5°C (34.7°F) through concerted climate action and realistic Nationally Determined Contributions (NDCs). NDCs are the individual country plans that need to become a reality to slow down the rate of heating.

3. Biodiversity Loss

The International Union for Conservation of Nature's (IUCN) Red List is a critical indicator of the health of the world's biodiversity. Currently there are more than 147,500 species on the IUCN Red List, with more than 41,000 species threatened with extinction.[11]

Deforestation, agriculture, urban sprawl, industry and pollution all destroy or damage habitats and contribute to the loss of plant and animal biodiversity.

Data compiled by the World Resources Institute reveals that our planet has already lost 80% of its forest cover to deforestation. The West Africa region, which boasted lush green tropical forests in the 19th century, has been stripped of 90% of its forest cover over the past 100 years. A similar trend of deforestation continues in the two remaining rainforest biomes in South America and parts of Asia.

According to the United Nations Framework Convention on Climate Change (UNFCCC), agriculture is the main cause of deforestation on the planet. Subsistence farming accounts for 46% of the total deforestation in the world, commercial agriculture 32%, logging 14% and fuel requirements 5%.

Why is the loss of these forests such an issue?

While tropical forests only cover about 7% of the Earth's dry landmass it is estimated they are home to half of all plant and animal species found on Earth. Many species within the rainforest are extremely vulnerable to extinction because they are specialised to microhabitats only found in those small areas.

Forests are also critical to our existence.

Over 40% of the world's oxygen is produced from rainforests. They contribute to the balance of gases such as oxygen, carbon dioxide, nitrogen and humidity in our air. Trees absorb carbon dioxide and play an important role in extracting the gas from our atmosphere.

Forest cover is vital for protecting watersheds (i.e. water collection basins). Deforestation leads to soil erosion and silting of our rivers, which impacts access to clean water.

Of extreme significance to our own survival is the genetic diversity of tropical forests. Cures for human diseases such as cancer may be contained in the genes of plants, animals, fungi and bacteria. These genes might also be key to improving yields and nutritional quality of foods which will be critical for feeding our exploding human population in coming decades. In the Amazon basin alone, more than 1,300 species of forest plants are used for medicinal or cultural purposes.[12] When a species is lost, it is lost forever. The extinction of one species also has a knock-on impact on the food chain which in turn upsets the entire ecosystem.

Our lives are inextricably linked with biodiversity and its protection is essential for our very survival.

Everyone has the human right to safe drinking water, yet currently, 785 million people worldwide do not have access to clean water.

4. Water Scarcity

Everyone has the human right to safe drinking water, yet currently, 785 million people worldwide do not have access to clean water.[13] This number is expected to increase substantially by 2050 when our global population approaches the projected 10 billion and demand for food and water increases.

Climate change, urban and industrial pollution, agricultural run-off and invasive species threaten the health of our rivers and groundwater systems. Overcoming the global crisis of water scarcity and insecurity requires deliberate prevention attempts rather than offsetting impacts once they arise.

Improved land management, irrigation techniques and placing an emphasis on protecting ecosystems will help secure water sources for generations to come.

5. Ocean System Collapse

Despite human habitation being largely confined to land, we've managed to significantly impact the health of our oceans and place them under immense pressure. As greenhouse gases in our atmosphere trap more energy from the sun, our oceans are absorbing more heat. The results of this are:

» coral bleaching events
» jeopardising the communities of marine life that depend upon coral reefs for shelter and food.

One serious impact of increasing levels of carbon dioxide in our atmosphere is ocean acidification. Around a quarter of the carbon dioxide produced by humans is absorbed by the oceans. As the carbon dioxide dissolves in seawater it forms a weak carbonic acid, making the ocean more acidic. The acidity of our oceans has increased approximately 30% over the industrial era.[14] This rate is 100 times faster than any change in ocean acidity in the last 20 million years. If our emissions of carbon dioxide remain relatively unchanged, scientists predict the ocean will become 150% more acidic by 2100.

It is widely believed that this increase in acidity is negatively impacting sea life, particularly corals, algae, crustaceans and molluscs. Creatures with structures made of calcium carbonate are especially susceptible as their shells and skeletons can weaken or dissolve in the acidic waters. Ocean acidification may also threaten plankton, which forms the base of the marine food chain and is the key to survival for larger fish and marine mammals. Impacts will be felt up the food chain and are expected to have a devastating impact on fish populations that are already under immense pressure from overfishing.

6. Pollution

Pollution is the introduction of a substance into the natural environment (land, water or air) which has harmful or poisonous effects.

Pollution can arise from:

- » natural processes such as erupting volcanoes or bushfires
- » anthropogenic sources (i.e. caused by humans).

Emissions from motor vehicles and heavy industry such as chemical plants, power stations and refineries are the leading sources of air pollution. Disposal of hazardous and municipal wastes to landfill are major sources of soil and groundwater contamination. Water pollution is predominantly caused by agricultural and urban run-off, industry discharge and sewage.

While attention in the past has focused on managing pollution from substances including heavy metals, dioxins, hydrocarbons and polychlorinated biphenyls (PCBS), emerging contaminants such as perfluoroalkyl and polyfluoroalkyl substances (PFAS) and microplastics will require a considered focus moving forward.

7. Waste

Waste is the term typically given to an unwanted or unusable material, substance or by-product. Every product has a life cycle that generally starts from raw material and ends in final disposal. Waste may be solid, liquid or gaseous and can be hazardous or non-hazardous. Waste is classified according to its source (municipal, commercial and industrial, construction and demolition) or by composition (organic, paper, glass, metal and plastic). Waste from obsolete electronic goods (e-waste) is one of the fastest growing types of waste.

In order to make a significant difference to waste generation, we need to do more than reduce, reuse and recycle.

Waste generation and disposal impacts our environment in several ways:

» The burial of waste in landfills contributes to soil, groundwater, surface water and even atmospheric contamination (from the release of gases from rotting organic waste).

» Waste can also have a significant impact on wildlife, particularly marine life.

Most people don't think about the types and quantities of waste they produce, yet waste is a growing problem. Over 2 billion tonnes of municipal solid waste are generated globally each year. This volume is projected to increase by roughly 70% to 3.4 billion metric tonnes by 2050[15] as the two drivers of waste generation—prosperity and urbanisation—continue to advance, particularly in developing countries.

In order to make a significant difference to waste generation, we need to do more than *reduce, reuse and recycle*. The key to reducing waste generation lies in avoiding wasteful purchases or activities in the first place and prolonging the lifespan of items that you do own. Developing a zero-waste mindset takes time and considerable, ongoing effort as it goes against the message of marketing campaigns of just about every consumer product on the market.

Why I wrote this book

Even as a young girl, it seemed so obvious to me: we've only got one planet and we need to treat it with the respect it deserves.

But I struggled to convince the people around me to care as much as I did. And I *really* struggled watching how society seemed to be ignoring the problem altogether.

Being just a wee bit passionate (!) I threw myself into studying environmental engineering and was determined to build a career that could help me create change on a grand scale. I completed my Bachelor of Environmental Engineering (with First Class Honours) in 1998 and then spent more than a decade monitoring and managing the environmental impact of some of the largest industrial sites in Australia, while completing a Master of Science in Environmental Geochemistry.

Me as a young environmental engineer at Hobart Zinc Smelter, 1999

My work has involved:

- » monitoring pristine streams on the boundaries of new mining projects
- » monitoring and managing water quality from tailings and waste rock storage facilities on operating or legacy mine sites
- » monitoring and managing dust and gas emissions from mines and smelters
- » testing degraded groundwater resources beneath municipal landfills and industrial sites
- » devising environmental management systems for industrial sites including mines, smelters and large sewerage treatment plants
- » monitoring dust, gas and radiation levels in our towns and cities.

This has given me the chance to witness firsthand the impact our modern lifestyle and insatiable demand for resources continually has on our environment.

My passion for helping families reduce their water and energy consumption developed during my time volunteering as a home sustainability auditor in my local community. While I'd always loved working within the beast driving change, I was starting to realise I could make a greater difference at the real 'coalface' in people's homes.

While pregnant with my first son in 2009, I noted the absence of information and products to help parents minimise the environmental impact of bringing a baby into the world. I created 'Sustainababy', an award-winning

I'm a firm believer that the biggest impact you can make, right now, is in your own household and local community.

eco-parenting resource, to guide parents in raising their babies with the environment in mind. I also developed several online sustainable living courses and launched my 'Eco Chat' podcast to share sustainability stories and practical information to help people live lighter. After six years of growing 'Sustainababy' I sold the business and returned to environmental engineering, focusing my efforts on supporting industry and governments to decarbonise and improve resilience to climate change risks.

My passion is sustainability and I love nothing more than inspiring and guiding people, particularly parents, to live lighter. I'm a firm believer that when it comes to the health of our planet, the biggest impact you can make, right now, is in your own household and local community.

That's the reason I've written this book; I want you to know the truth about the state of our earth, but I don't want you to feel so overwhelmed that you believe it's all too hard and you can't make a difference, because that couldn't be further from the truth.

Since you've made it this far, I suspect you're ready to make a change for the better. Perhaps you're ready to end your love affair with fossil fuels and embrace renewable technologies. Maybe you're keen to ditch toxins in your home or break up with single-use plastics. Or you might simply want to grow something in your garden that you can turn into a family meal without the food miles.

If any of this sounds like you, you're in the right place and I'm so glad you're here.

Whether you're a lifelong greenie looking for some extra inspiration or someone who's only recently developed a passion for conserving our environment, I'm positive this book will help you focus on the many small changes you can make to drive positive environmental change in our world.

This book is divided into three parts:

» **Part 1** sets the stage for sustainable success.

» **Part 2** covers my seven pillars of sustainable living to help you reduce your impact.

» **Part 3** describes how you can expand your impact by raising eco-conscious kids, voting for and investing in the future.

Each chapter revolves around a central argument: *You have so much more power than you realise, and meaningful change can be accomplished through the tiniest of actions.*

Let's get started.

PART 1

SET THE STAGE FOR SUSTAINABLE SUCCESS

CHAPTER 1
Move beyond climate grief

"You can't unhear the truth."

——Laura Trotta

Public awareness of global warming seemingly changed overnight with the release of the 2006 documentary *An Inconvenient Truth*. The film documented former United States Vice President Al Gore's campaign to educate people about global warming. And educate it did. The film raked in USD53 million at the box office worldwide.[16]

Regardless of whether you have viewed the film or not, you have heard the Truth about the state of our precious earth. Just the fact you're reading this book indicates you appreciate that an understanding of the Truth, however inconvenient it may be, comes with a responsibility to act.

But how can you act if you're struggling to process the Truth?

How can you move forward if you're overcome by eco-anxiety or climate grief?

My first real appreciation of the paralysing nature of climate grief occurred after I trained with Al Gore to become a Climate Reality Leader. Following completion of the training, leaders are expected to regularly log climate leadership acts such as:

» delivering a public presentation on climate change

» writing to a member of parliament for climate action

» writing a blog or recording a podcast on the topic.

A few months after my training, I attended a meet up of climate reality leaders in my home city of Adelaide. It was here where another leader shared quietly with me that in the two years since her training, she hadn't undertaken one climate leadership act. And it wasn't because she didn't want to. She simply wasn't up to it. She'd become so overwhelmed with the enormity of the climate crisis that just managing the demands of her profession as a doctor in general practice was a challenge. Her despair was made greater by the fact she'd been beating herself up for not doing more. Sadly, she isn't alone.

A 2019 poll by the American Psychological Association revealed that 68% of US adults are experiencing at least a little anxiety about climate change and almost half (47%) of young respondents aged 18 to 34 said their anxiety about the climate was affecting their daily lives.[17] I would suspect that a similar poll in Australia following the 2019/2020 Black Summer bushfires and 2022 Queensland and New South Wales floods would yield higher percentages again.

What does climate grief look like?

When we experience a loss strongly enough, the natural human response is to grieve. Many of us are grieving, or will grieve, a changing climate and the resulting ecological losses. Climate grief is ecological grief and represents a loss (or anticipated loss) of something we value in our local or global environment. Along with eco-anxiety, the term climate grief is increasingly used to describe a common and pervasive psychological response to the ecological crisis, and to climate disruption in particular.[18] Psychologists use the term climate grief to refer to feelings of sadness, loss and anxiety in response to climate devastation.[19] This grief may be felt in relation to experienced or anticipated ecological losses, including the loss of species, ecosystems, and meaningful landscapes due to acute or chronic environmental change.[20]

Just like with the loss of a loved one, those of us facing climate grief will move through the Kubler-Ross five stages of grief.

In February 2017 the Australian Treasurer and future Prime Minister brought a lump of coal into the House of Representatives and delivered a climate denial speech.

Denial

It was a performance worthy of an Academy Award.

In February 2017 the Australian Treasurer and future Prime Minister Scott Morrison brought a lump of coal into the House of Representatives and delivered a climate denial speech that started with:

"This is coal. Don't be afraid. Don't be scared. It won't hurt you. It's coal ..."

His intention was to strike fear in the hearts of the voting public by suggesting if the opposition were elected, Australia would become like South Australia and struggle to keep the power on.

Just five months earlier in September 2016, South Australia experienced a multi-day statewide power outage, brought on by an extreme weather event that damaged transmission and power distribution assets. The power outage occurred just four months after the state turned off its last coal power generator. Renewables were blamed for the blackout despite the actual cause being storm damage and overly sensitive trip switches. A smaller blackout in South Australia six months later was all Scott Morrison needed to bring the lump of coal into parliament and argue that South Australia's renewables transition was "switching off jobs, switching off lights and (sic) air conditioners and forcing Australian families to boil in the dark as a result of their Dark Age policies."[21]

The Federal Coalition Government didn't hold back—publicly ridiculing the South Australian Labor government and trying to seed doubt throughout the country as to the reliability of renewable technologies. To get an idea of some of the mud-slinging that occurred watch 'SA Premier Jay Weatherill shirtfront Josh Frydenberg' on YouTube for a good laugh.

The Australian public watched the political circus of climate change denial playing out amongst their elected leaders.

But they weren't the only ones watching.

The shenanigans caught the attention of Tesla's CEO Elon Musk, who famously made a Twitter bet with Aussie tech billionaire and Atlassian co-founder Mike Cannon-Brookes that Tesla would build a battery for the South Australian Government to reduce the risk of blackouts in 100 days of the contract being signed or the AUD 50 million system would be free.

The rest is history.

Tesla installed the world's biggest lithium-ion battery in South Australia within 70 days of Musk's promise and Mike Cannon-Brookes said he'd never "been happier to lose a bet."

But most importantly, the battery reduced the cost of operating the once shaky South Australian power grid by 91%, stabilised the electricity grid, facilitated integration of renewable energy and reduced the chance of load-shedding events. In its first two years of operation, South Australia's 'Big Battery' saved South Australian consumers over AUD 150 million.[22]

Mike Cannon-Brookes 🏗️🔔... ✔ @mcannonbrook... · Mar 9, 2017 ···
Holy s#%t

afr.com
Tesla battery boss: We can solve SA's power woes in 100 days
The head of Tesla's battery division says the company could solve SA's
power woes within 100 days, and do the same for Victoria.

💬 28   ↻ 497   ♡ 984   ⬆

Mike Cannon-Brookes 🏗️🔔... ✔ @mcannonbrook... · Mar 9, 2017 ···
Lyndon & @elonmusk - how serious are you about this bet? If I can make
the $ happen (& politics), can you guarantee the 100MW in 100 days?

  🔵 Mike Cannon-Brookes 🏗️... ✔ @mcannonbrook... · Mar 9, 2017
  Holy s#%t afr.com/news/tesla-bat...

💬 64   ↻ 673   ♡ 1,497   ⬆

Elon Musk ✔ @elonmusk · Mar 10, 2017   ···
Tesla will get the system installed and working 100 days from contract
signature or it is free. That serious enough for you?

💬 587   ↻ 7,465   ♡ 14.7K   ⬆

Mike Cannon-Brookes 🏗️🔔📧 ✔   ···
@mcannonbrookes

Replying to @elonmusk

legend! ☀ You're on mate. Give me 7 days to try sort
out politics & funding. DM me a quote for approx
100MW cost - mates rates!

4:16 PM · Mar 10, 2017 · Tweetbot for iOS

South Australia is now a world leader in renewables deployment. The State is on track to reach 100% renewables by 2030.

Mike Cannon-Brookes 🧢 🌊 📧 ✅
@mcannonbrookes   ...

⚡ Thank you @elonmusk, Tesla's amazing Aussie team, @jayweatherill & all SA 🙌 Never been more happy to lose a bet. 3x bigger than any 🔋 in world! Huge step for Australia & proving what we can do. Only lumps of coal req'd are for #AusPol🎄 stockings 🎉

fortune.com
Elon Musk's New Battery Just Won Him a $50 Million Bet
Musk bet he would be able to provide the world's largest lithium-ion battery within 100 days.

6:19 AM · Nov 24, 2017 · Tweetbot for iOS

South Australia is now a world leader in renewables deployment. The State is on track to reach 80% renewables by 2025, 100% by 2030 and 500% of current demand renewables by 2050.[23] The high-voltage electricity transmission interconnector between South Australia and NSW is expected to be fully operational by 2024, enabling South Australia to export renewable energy to the populated eastern states which are still heavily powered by coal. Foreign and local investment is pouring into the State, with multiple exciting green hydrogen and ammonia projects in the pipeline.

South Australians certainly aren't living in the Dark Ages.

Ever since scientists started warning the world about climate change, climate change deniers have been a huge barrier to solutions. Denial is the first stage of grief. Who knows if the denial prevalent in government leadership positions was caused by grief from the climate crisis or grief at losing a lucrative export coal industry? Regardless, this denial has cost us precious and critical time, well over a decade, to act and transition our national grid to renewables.

Thankfully the number of climate change deniers is reducing. Many were silenced or sent packing from government in the 2022 Australian federal election, due to the Australian public overwhelmingly voting for climate action. But they're also reducing in number due to the compounding scientific evidence of climate change impacts and trailblazing jurisdictions like South Australia leading the way with decarbonising their grids and paving the way for other states to follow.

Since you're reading this book, I'm assuming that you've passed, or even bypassed, the denial stage.

Anger

The younger generation, led by Greta Thunberg and her 'Fridays for Future' movement, have demonstrated how climate grief can look like anger. Their anger is a justified reaction to the climate denial and failure of older generations and global leaders. But they haven't let themselves get stuck in feeling angry. They have harnessed their anger and used it to fuel meaningful action that has enabled them to be heard.

Bargaining

For many years corporations and governments have been trying to bargain their way out of climate action and good environmental management:

» If they plant a forest of trees and offset their carbon emissions, can they continue their polluting activities unchanged?

» If they revegetate an area of landscape and reintroduce native species, can they destroy a different parcel of land to extract the resources beneath?

» If they train and employ members of a local Indigenous community, can they destroy a cultural site that's in the way of extracting valuable resources?

The business community might formally call the above activities 'offsetting', but they're all essentially examples of bargaining.

Bargaining can take on all sorts of forms on an individual level too. For example, if we take our reusable shopping bags to the supermarket can we accept the free supermarket collectable toy for our kids? If we buy a hybrid car, can we then drive to work instead of catching public transport?

Bargaining has a place in getting us to take those first small steps to becoming more sustainable. But it's important not to

Reaching the 'acceptance' stage of climate grief includes coming to terms with the truth about the climate crisis.

get stuck in the bargaining phase as it will get in the way of some of the bigger steps needed to deliver action.

Depression

When climate grief becomes too heavy, people can struggle with feelings of depression and overwhelm that can cause them to shut down. The climate reality leader doctor friend I mentioned earlier was likely in this stage of grief following her training. Others may also become so involved and overextended in their sustainability work or climate activism that they burn out.

Acceptance

Reaching the 'acceptance' stage of climate grief includes coming to terms with the truth about the climate crisis, calmly accepting the facts and processing your complex feelings while remaining functional and engaged with the world. In this way you're able to stay motivated and focus on actions that are meaningful and sustainable, without becoming depressed or withdrawn. It doesn't mean that you're happy with the situation, just that you're able to move forward in a productive way.

Moving through climate grief

That's all well and good I hear you say, but how do I move through this intense anger I'm feeling at the inaction of world leaders, or my rage at older and previous generations that have caused this destruction, or at energy companies still pushing to develop new coal mines and gas plants? How do I move through the intense feeling of sadness and despair every time I watch the news and see the latest destruction from an extreme weather event?

Time is a great healer, but sadly we don't have the luxury of time to wait for much of the population to grieve climate change and the resulting ecological fallout. We need to collectively accept the science, accept the damage that has occurred and work together calmly and quickly to adapt, build climate resilient communities and minimise future ecological damage.

I offer the following strategies to help you move through your climate grief. These have helped me manage my emotions over the years and stay the course.

1. Acknowledge your grief

We were on our annual family holiday on South Australia's beautiful Limestone Coast in December 2019 when much of the country went up in flames in the Black Summer bushfires. Like many Australians I was glued to the news, watching in horror as Victoria's East Gippsland, the holiday location of my childhood, burnt from the mountains to the coastline. The Royal Australian Navy was needed to evacuate Mallacoota's residents by sea.

Despite being safe and sound with no immediate risk to myself, I was absorbed in bushfire maps and warnings for many regions in Australia and was not present with my family.

I was grieving for the thousands of hectares lost to the fire, the millions of animals that were perishing in the flames, the tens of thousands of Australians who lost their homes or businesses and who had their annual summer holiday ruined or worst still, lost their life. If there was a medal for grieving on behalf of future generations who wouldn't enjoy the carefree summers I enjoyed in my childhood, I'd have won that too!

I couldn't hide my grief from my family, so I confronted it, told my husband exactly why I was upset and had a really good cry. Just acknowledging what I was feeling helped me pull myself out of despair so I could redirect my energy into creating family holiday memories for my kids and prioritising my rest so I could manage another intense year working in the sustainability field.

If you're struggling with climate grief the best thing you can do is to acknowledge it's there. Give it a name (Gertie Grief is back, back again!). Recognise when you're having a sad day. Then turn off the news and social media and do something kind for yourself. If you're still struggling, lean into your community of like-minded friends for support or take some climate action to be part of the solution (more on that soon!).

THERE ARE A FEW ONLINE COMMUNITIES THAT CAN HELP YOU FEEL LESS ALONE INCLUDING:

» My Sustainability in the Suburbs group on Facebook

» The free Facebook community Australian Parents for Climate Action

(You can find both of these simply by searching their names on Facebook.)

2. Recognise you're not alone

In March 2020, when the Australian Government closed international borders and recommended that everyone stay at home for six weeks to 'flatten the curve' of the COVID-19 virus, Australian lives and livelihoods were disrupted in a way they hadn't been since World War II. But somehow, through the anxieties of lost income and the unknown of what lay ahead, a large portion of the population were baking bread and catching up on projects around the home to the soundtrack of Ben Lee's *We're All in this Together*. The knowledge that we indeed *were* all in this together was a source of comfort.

Similar to the COVID-19 pandemic, not one person on Earth will be immune to the impacts of climate change. If you're struggling with climate grief, recognising that others are too may provide some comfort.

3. Make progress visible

Last year I was invited to share a message of innovation and progress at a Doctors for the Environment Australia conference. I delivered a presentation on the engineering solutions to climate change with the goal of giving the several hundred doctors in attendance a sense of hope for our future. What struck me during my presentation was just how much the audience needed this message of hope.

Medical doctors are already dealing with the health impacts of climate change every day in their practices.

Health impacts like:

» an increase in asthma from bushfire smoke and thunderstorms

» heatstroke deaths during heatwaves

» the spread of infectious diseases during flood events

» new vector-borne diseases and resulting pandemics (hello COVID-19!)

» a significant increase in mental illness.

They're also counselling couples who are choosing not to have children due to the climate crisis. Doctors rightly believe that climate change is a health issue and they are deeply concerned.

Through the process of preparing and delivering my conference presentation, I became acutely aware of how my sustainability work balances out my own climate grief. Barely a day passes when I don't hear of plans for a new solar, wind, battery or hydrogen project from my renewable engineering or environmental planning colleagues. If I'm not actively working on a project to support an organisation to decarbonise, I'm preparing a tender to do so.

The climate risk and resilience work to help businesses and governments adapt to the impacts of climate change can be quite confronting. But seeing firsthand the speed of action being taken by the business community to decarbonise is exciting and uplifting. Most engineers and scientists I work with appreciate how privileged they are to be working in such an historical time and they are giving their all to help society transition to net zero.

Noting the above has made me realise: **you can't see what you can't see.**

If you're not directly working in this space, your source of progress on climate action is likely the constant news stream which is dominated by another extreme weather event, latest climate prediction or political delaying tactic. These things just fuel more climate grief.

I want you to see this progress so you can feel positive too.

Even if you can't directly see the action of the engineers, scientists and entrepreneurs working tirelessly on solutions to climate change, I want you to know it's happening. Businesses, industries and organisations (including many government organisations) are moving ahead at lightning speed to decarbonise. They will continue to do so with or without solid federal and state net zero targets because:

- » Their shareholders, communities and employees are demanding it.
- » It's required by their insurers and financiers.
- » Their business will be left behind, or at worse, fail, if they don't.
- » They won't attract or retain talent.
- » It's the right thing to do.

How can you make the above visible if you're not working in the field of climate change solutions? Simply follow the

It's only when we become part of the solution that we can move through anger and depression to acceptance.

social media channels of renewable technology firms like ACCIONA and engineering and sustainability consulting firms like GHD, WSP, ERM, AECOM, Arup and Aurecon. Before you know it, your newsfeed will soon fill up with active projects to overcome the climate crisis and design resilient communities. If that's not good medicine, I don't know what is!

4. Find your way of taking action

For many of us, the best way to manage our eco-anxiety and climate grief is to act. It's only when we become part of the solution that we can move through anger and depression to acceptance. For you, taking action may include:

- » joining a climate activist organisation
- » pivoting your career into a sustainability field
- » actively supporting members of parliament who champion environmental issues
- » simply reducing the environmental footprint of your own household.

Final thoughts

Climate grief and eco-anxiety are not only real afflictions, they're on the rise, impacting a growing percentage of our global population. By recognising your eco-anxiety and moving through climate grief, you'll be more able to turn the paralysis that can often accompany grief into positive action and thrive in the process.

Take action now!

It's over to you to turn any grief you're feeling over the ecological and climate crisis into positive action. You can start by taking the following three steps:

1. Take a moment to reflect on where you are in the Kubler-Ross stages of grief in relation to the climate crisis. Are you in denial, anger, bargaining, depression or acceptance? You don't need to do anything with this awareness, just acknowledge where you are.

2. Join a community of like-minded eco-conscious people taking action to improve the environment. First look to your local community for organisations you can join. A community garden, Landcare group or 'Friends of' group for your local natural wonder are always willing to welcome new members. If you'd love the support from an online community of eco-conscious people reducing their own environmental footprint, I'd love to welcome you into my Sustainability in the Suburbs community at **sustainabilityinthesuburbs.com.**

3. Make the progress of climate change solutions visible to you. Follow the social media channels of renewable technology firms and engineering and sustainability consulting firms to fill your newsfeed with sustainability positivity.

Share what action you've taken to move beyond climate grief on social media with the hashtag #sustainabilityinthesuburbs and tag me @lauratrottadotcom so I can personally congratulate you.

CHAPTER 2
Practise sustainable self-care

"*You can do anything, but not everything.*"
—DAVID ALLEN

There have been times in my life when I've maintained a great balance between my work, hobbies and family responsibilities. But sadly, there have been more times in my life where my passion for the environment has got the better of me; I've worked myself to the bone and I've burnt myself out. These days I juggle my work and life commitments with an autoimmune condition brought on by chronic stress and overwork when I was building a sustainability e-commerce business around providing full-time care to my baby and toddler. During that period, I was trying to do everything: be CEO, earth mother and perfect wife all at the same time, sacrificing my sleep and rest in the process. Thankfully, with the support of medical specialists, I was able to bring myself back from burnout. These days I'm ruthless in my self-care to ensure I never experience burnout again.

Acting on climate and ecological issues can be incredibly fulfilling. It can also be exhausting and consuming—there is always something that could be done faster or better. If you're at the early stage of your eco journey, you're at greater risk of overwhelm between the shock and realisation of the severity of the crisis gripping our planet, and the seemingly endless actions you can see you need to take to reduce your own environmental footprint.

Before getting into practical tips for sustainable living I want you to understand that you don't need to do everything. And you sure as hell don't need to do everything at once. You're no good to anyone, let alone the sustainability cause if you're exhausted, burnt out and a shell of your former self.

When you prioritise your self-care, you're much more likely to reach your end goal, even if it does take a little longer than you'd hoped to get there. You're more likely to enjoy yourself along the way (isn't that what life is about)? And most importantly, you create a ripple effect, inspiring others to prioritise their self-care too.

Here are my 10 best self-care tips to ensure your eco journey is sustainable and enjoyable for you and your family.

1. Ditch perfectionism

Self-compassion is the antidote to perfectionism. Through self-compassion, we give ourselves permission to be human. Perfection is not sustainable, so please accept and love yourself for who you are right now. Regardless of how many sustainable changes you do or don't make, or how much climate action you do or don't do, you are enough, just as you are.

2. Spend time in nature

Spending time in nature is so good for you. It promotes happiness by increasing endorphin levels and dopamine production, reduces symptoms of anxiety and depression, and lowers cortisol (the stress hormone). Time in nature also fosters our connection to Mother Earth.

How much time in nature should we aim for?

Rachel Hopman, Ph.D., a neuroscientist at Northeastern University recommends we live by the 20-5-3 rule[24]:

» Spend 20 minutes, three times a week outside in nature (like a neighbourhood park, suburban beach or botanical garden) to boost cognition and memory, reduce the stress hormone cortisol and improve feelings of wellbeing.

» Spend a minimum of 5 hours a month in semi-wild nature, (like a national park), to improve happiness, reduce stress and give your body a break from the frenetic pace, loud noises, rotten

Cultivating a meditation practice has helped me to calm my racing mind, focus my attention and reduce stress.

smells, pinging phones and to-do lists of urban life.

» Spend 3 days per year off grid in nature, camping or renting a cabin (with friends or solo) in a location with little or no mobile phone reception, plenty of wild animals, away from the hustle and bustle of urban life. This retreat will cause your brain to ride alpha waves, the same waves that increase during meditation or when you lapse into a flow state. They can reset your thinking, boost creativity, tame burnout and just make you feel better.

3. Learn to relax

It can feel like such a waste of time being idle when there are a thousand things to do and a climate crisis to solve. But our bodies and minds need downtime to renew. Time spent in renewal is not wasted. It is R & R for your mind, body and soul.

For over 20 years, my regular yoga practice has helped me navigate some of the most intense periods of my life. I find Restorative Yin Yoga the best to calm my nervous system. Similarly, cultivating a meditation practice has helped me to calm my racing mind, focus my attention and reduce stress.

Magnesium is essential for the healthy functioning of muscles and for the management of protein, bone health, blood sugar levels and blood pressure. Our body's magnesium reserves are easily depleted under stress but can just as easily be replenished by soaking in a magnesium salt bath. I add a cup of magnesium chloride flakes to a warm bath each week and soak for an hour to ease any sore muscles and promote relaxation. Don't have a bath? Soaking your feet in a bucket of magnesium salts will have a similar impact.

It's not important how you relax, just that you do relax. Whether yoga and meditation are your relaxing go-tos, or you prefer to soak in a bath or curl up on the couch with a good book, the important message here is to prioritise downtime so you're here for the long time.

4. Reduce screen time

I loved my analogue childhood, playing outside with my sisters back in the 1980s. It was a time where dinner rather than a screen was the thing that would finally entice us back inside, exhausted but happy after a long day of play. The world has changed significantly since then and, like many of you, I'm working and parenting in the digital age. Too much screen time can leave both our kids and ourselves wired and tired. This impacts on our connection time with family and friends and contributes to poor sleep, eye strain and headaches, addiction, muscle aches and sedentary behaviour.

How much screen time is too much?

Most experts agree that adults should limit screen time to less than two hours per day outside of work-related activities.[25] Mindful acts like the following can help you manage your screen time and not have it manage you:

- » making mealtimes and bedrooms a screen-free zone
- » turning off notifications
- » setting screen time limits
- » tracking your time and taking breaks to stretch and move around every 30 minutes while using a computer.

In addition to reducing screen time overall, consciously reducing screen time in the evenings helps to ensure a good night's sleep. Until the advent of artificial lighting, the sun was the major source of lighting, and people spent their evenings in (relative) darkness. The proliferation of electronics with screens, as well as energy-efficient lighting, is increasing our exposure to blue wavelengths, especially after sundown. Blue light can affect sleep and potentially contribute to the causation of cancer, diabetes, heart disease and obesity.[26] Blue light throws the body's biological clock—the circadian rhythm—out of whack, blocking a hormone called melatonin and interfering with our body's ability to prepare for sleep. To combat this, avoid looking at screens two to three hours before bedtime and especially 30 minutes prior to sleep.

5. Get moving

Moving our bodies is essential for our wellbeing. Regular exercise increases self-confidence, improves our mood, helps us relax and lowers symptoms of mild depression and anxiety. Exercise can also improve sleep, which is often disrupted by stress, depression and anxiety.

Sleep is a critical component of the self-care puzzle that allows our minds and bodies to recharge.

For most healthy adults, the Department of Health and Human Services recommends getting at least 150 minutes of moderate aerobic activity or 75 minutes of vigorous aerobic activity a week, (or a combination of both).[27] I walk along my local beach a couple of times a week (getting my dose of nature in as well) and enjoy group training sessions with a local personal trainer three times a week to build muscle and connect with friends at the same time.

My advice is to find movement you enjoy and commit to it regularly. Last time I checked, walking was free!

6. Prioritise sleep

Sleep is a critical component of the self-care puzzle that allows our minds and bodies to recharge. In fact, it's one of the three activities we humans must do to survive: eat, drink, sleep, repeat. Good sleep also helps the body remain healthy and stave off diseases. Unfortunately, sleep is often the first thing we're willing to compromise when we're busy. It's also something that can suffer when we're stressed or anxious.

How much sleep is enough?

Most adults require between seven and nine hours of sleep per night for proper cognitive and behavioural functions.[28] It appears that when you nod off is just as important as how long you stay asleep. A study by the United Kingdom's University of Exeter found that falling asleep at 10 pm or shortly after is associated with a lower risk of developing cardiovascular disease compared with falling asleep earlier or later at night.[29]

7. Eat well

Our energy levels and general health and wellbeing are highly influenced by the quality of our diet. It's easy to feel confused as to exactly what a quality diet is given all the different messaging and fads out there. In the end, just keep things simple: eat real food, locally produced, in season and organic wherever possible. Minimise processed foods and sugar. It shouldn't come as a surprise that the diet that's best for our bodies is also best for our environment (more on that later!).

8. Connect with friends

If the COVID-19 pandemic has taught us anything, it's that we're social creatures and connecting with each other in person is fundamental to our happiness. Make time for friendships and the relationships that matter. If you catch up with your friends in the great outdoors or exercise together, you'll tick two things off your self-care list at the same time!

9. Make time for fun/hobbies

Hobbies are often thought of as activities for people who lead quiet, relaxed lives. In reality, it's people with busy and stressful lives who may need hobbies even more than the average person.[30] Hobbies are a great outlet to immerse yourself in an activity you love or are passionate about.

Many of my friends and colleagues ask how I find time to play trumpet in a number of Adelaide bands when I'm so busy with my family and work. The answer is simply that I make the time. I'm happiest when I'm performing in a band, providing the audience and myself with a brief interlude to the mundane or serious aspects of life. And my family are happier when I'm happy.

Me playing in the horn section of Adelaide ska band, The Overits

If you've let a hobby you love fall by the wayside, I encourage you to block time out in your calendar to reacquaint yourself with your passion.

10. Practise gratitude

The simple act of expressing thanks for our lives is so beneficial in times of challenge and change. It's also great for our health. People with an 'attitude of gratitude' experience lower levels of stress and increased levels of physical and mental wellbeing. Gratitude stems from the recognition

I can't stress the importance of looking after yourself so you can make sustainable lifestyle changes easily.

that something good happened to you, accompanied by an appraisal that someone, whether another individual or an impersonal source, such as nature or a divine entity, was responsible for it.[31]

How you choose to practise gratitude is up to you. It may be through journalling, prayer or undertaking random acts of kindness for strangers (without videoing the act and sharing on social media for validation). If you're new to practising gratitude, perhaps start by taking 5-10 minutes to write down a list of all the things you're grateful for in your life. Choose one item from your list and reach out to the person who was wholly or partially responsible to express your thanks and gratitude for them. Take note of how you feel afterwards.

Final thoughts

I want this book to be a practical handbook for you to make your home and lifestyle more sustainable. But I can't stress the importance of looking after yourself so you can make sustainable lifestyle changes easily, without adding more stress onto your overflowing plate. By creating sustainable self-care practices, you'll increase your positivity and energy levels and be much better placed to reach your green goals.

Take action now!

In Part 2 of this book, I share actionable strategies to help you create a healthier, more sustainable home. But before we dive in, let's ensure you've locked in some sustainable self-care practices by completing the following three actions.

1. Set an alarm reminder on your phone right now for 9:30 pm each evening to remind yourself to turn off all screens, put your smartphone in airplane mode and start heading off to bed so you can fall asleep within the miracle hour of 10-11 pm.

2. Block time in your calendar now to spend 20 minutes each week outside in nature and 5 hours each month in semi-wild nature (a monthly walk and picnic in a national park is perfect!). For bonus points book a holiday in the next year to spend a minimum of 3 days off grid in wild nature. Glamping is totally allowed, as long as you switch off your phone!

3. Write a list of what you are grateful for in your life today. Reach out to someone who is responsible or partially responsible for one item on your list and express your thanks and gratitude to them. For bonus points call them directly or write them a card or letter of thanks and post it in the mail. You will make their day.

Share what changes you've made to improve your self-care on social media with the hashtag #sustainabilityinthesuburbs and tag me @lauratrottadotcom so I can personally congratulate you.

PART 2

REDUCE YOUR IMPACT

CHAPTER 3
Embrace ecotarianism

> *"To eat is a necessity, but to eat intelligently is an art."*
> —FRANCOIS DE LA ROCHEFOUCAULD

My journey to a sustainable home and lifestyle started with an overhaul of my diet in my early twenties. I'd love to tell you this was a conscious choice, but in reality, it was driven by health issues.

I'd always believed I ate healthily: cereal, porridge or baked beans for breakfast, a salad roll and apple for lunch, and meat and three veg for dinner. Even on university trips where the bus would stop at a fast-food outlet for everyone to fuel up, I'd find a local bakery to purchase a salad roll.

But I was also a young woman of the early 2000s when fat was the enemy and diet products were king. Diet cola was my vice and foods like diet yoghurt and sugar-free chewing gum were staples.

I was working long hours as a fly-in fly-out environmental engineer at a remote mine site. As I didn't drink coffee, I turned to diet cola to give me the early morning jump start to get through the 12 hour shifts ahead (14 of them in a row). I'd often open my first can at 5:45 am on the way to work, and then I'd have another after lunch to push me through the afternoon.

My wakeup call came after a trip to the dentist. My teeth had become sensitive and eating cold foods became painful. The dentist revealed that the enamel on my two front teeth had worn away and I needed a filling in each. It turns out that the acidity of the diet cola had dissolved my enamel. The dentist's advice was to drink my diet cola with a straw to minimise contact with my teeth.

I thought if the diet cola is doing this to my teeth, what impact is it having on my overall health?

As it turns out, I didn't need to look too far.

I was also suffering from chronic migraines, the type that would be accompanied by an aura, excruciating nausea and made me retreat to a cold, dark room, often for days on end to recover. These were becoming more frequent, popping up on most weekends. When they started coming on during work hours, I realised I needed to take serious action.

Following the advice of my doctor, I started keeping a food diary, researching food triggers and learning the numbers of the additives and preservatives that gave me the most grief. It was no surprise that the artificial sweetener aspartame (additive 951) came up top as a trigger for my migraines. I went cold turkey on the diet cola, suffering an awful week of withdrawal headaches, and haven't touched the stuff since. That was 20 years ago and I'm proud to say my migraines are well under control and now a rare occurrence.

Food and drink are intertwined throughout every aspect of our culture and it's not always easy to make good choices in this area; choices that are good for our own health as well as for our planet. Think of any religious or life celebration (hello birthday cake!!) and you can bet your bottom dollar that food plays an integral role. Cooking shows also dominate the top TV ratings, #foodporn dominates our social media feeds and many of us are contributing to this trend by posting more pics of the meals we're cooking and eating than of our family and friends!

It's true that we are what we eat but even more so, our world reflects what we eat. We all need to become more mindful of what we eat as our global human population continues to climb.

With so many people struggling to eat a healthy diet (the high rates of obesity and diabetes speak for themselves) it can be hard to wrap your head around the concept of sustainable eating and its importance. But I can't stress its importance enough. The very survival of our species depends on us changing the way we eat. With the world's human population predicted to increase over 30% from its current level of 7.5 billion to 10 billion people by 2050, it's one of

The very survival of our species depends on us changing the way we eat.

the most inconvenient truths that needs to be addressed ... and quickly.

Agriculture is the largest land use on Earth. With more and more wilderness areas like rainforests being transformed into plantations, farms or just grazing land for livestock, we could literally run out of space to grow the food we need to survive. And we will continue to witness massive losses of biodiversity as our impact on this planet increases with our growing population.

But it's not just changing land use and subsequent loss of biodiversity that goes hand in hand with a growing global human population:

» Intensive agricultural practices together with synthetic fertilisers and pesticides degrade our soils and impact the health of workers and nearby communities.

» Livestock are the number 1 emitter of greenhouse gases (and therefore climate change!).

» Large amounts of land in developing countries are being cleared to grow feed for factory-farmed animals in the developed world.

» There is a growing reliance on genetically modified crops.

» More than one third of our oceans are over-fished.

If these are an example of current impacts of agriculture, what will the environmental impacts of our diet be when our global population reaches 10 billion people?

We can continue to eat ourselves out of house and home OR we can eat more sustainably and give current and future generations the gift of a good quality of life and natural environment.

The writing's on the wall. Our planet cannot easily sustain a human population of 10 billion at current consumption levels. It's time to collectively change the way we live, and the best place to start is by changing the way we eat.

A century ago, farm animals were only seen grazing in the fields or even in people's backyards. Nowadays unmarked sheds on the outskirts of large cities house animals for the meat market and every day in the developing world, forests are felled to grow feed for these factory farms.

A century ago, people mainly ate food that came directly from a plant or animal. Nowadays our supermarket shelves are lined with processed foods and drinks. These foods (and I use the term 'food' loosely here) have infiltrated our homes and schools and our health is suffering as a result.

Like all aspects of sustainability, what's good for the environment is also good for our health. Cleaning up your diet by reducing consumption of processed foods and eating more wholefoods and plant-based foods is one of the best actions you can take to improve your health and the health of our planet.

Wholefoods have gained in popularity over the past decade, and so they should—they're amazingly good for us. But we do need to take it one small step further in our journey towards sustainability.

We need to embrace eating sustainably.

We need to embrace *ecotarianism*.

In this chapter I'll share how you can become an ecotarian, but first, let's look at the benefits of sustainable eating.

Benefits of sustainable eating

Reducing your consumption of animal products and processed foods is essential when reducing your environmental footprint. This does require commitment and can take some time (as there's a degree of planning and preparation involved when you're making any changes to your diet) but the benefits are so worth it.

1. Eating sustainably reduces our global carbon emissions and helps to reduce the impact of climate change

In Australia, direct livestock emissions account for about 70% of greenhouse gas emissions by the agricultural sector and 11% of total national greenhouse gas emissions. This makes Australia's livestock the third largest source of greenhouse gas emissions after the energy and transport sectors.[32] Livestock such as cattle and sheep release methane via belching when they're digesting their food. Pigs and poultry produce small amounts of methane as the result of the incidental fermentation that takes place during digestion. Methane is a particularly potent greenhouse gas and is 25 times more effective as carbon dioxide at trapping heat in the atmosphere.

By reducing your meat consumption overall you'll be able to reduce your grocery bill.

By reducing our consumption of meat, we're reducing the demand for livestock and, therefore, emissions from this sector.

2. Eating sustainably will save you money

The price of meat has skyrocketed in the last few years in Australia, due to the impact of floods and bushfires on production and the pressure on supply chains during the COVID-19 pandemic. Some cuts of beef are over $50 per kilogram. The average cut of steak is hovering around $30 per kilogram, and that's not even an organic variety. Even beef mince, a common base for an economic family meal, is currently around $15 per kilogram.

By reducing your meat consumption overall you'll be able to reduce your grocery bill or at least offset the growing prices of fresh fruit and vegetables. (Fresh produce prices have been impacted by extreme weather events and supply chain issues during the pandemic.)

3. Eating sustainably boosts your immunity and general health

The Western Pattern Diet (WPD) is a modern-day style diet that mostly contains high amounts of processed foods, red meat, high-fat dairy products, high-sugar foods, and pre-packaged foods, that increase the risk of chronic illness. Chronic illnesses linked to WPD include obesity, diabetes, cardiovascular diseases, cancer, autoimmune diseases and Alzheimer's disease, which are rare or virtually absent in hunter-gatherers and other non-westernised populations.[33]

By cleaning up your diet and eating more wholefoods, you'll be better placed to maintain a healthy weight, improve your longevity and increase your energy levels and general quality of life.

What is sustainable eating?

Sustainable eating is about enjoying a healthy, balanced diet that's in tune with the environment, while still enabling Earth to easily provide for future generations.

Sustainable eating involves:

» reducing your consumption of processed and packaged foods

» reducing meat consumption

» increasing the portion of wholefoods in your diet, preferably sourcing local, in-season produce and preferably organic, where possible and where your budget allows.

But there's a little more to it as to eat sustainably, we also need to be mindful of food packaging and food miles and the answer isn't always straightforward.

Wholefoods are foods that have been processed or refined as little as possible and are free from additives or other artificial ingredients. Examples are brown rice, raw nuts, meat, eggs, fruit and vegetables. Wholefoods have grown in popularity in recent times and for very good reason. They're 'real' food and are the traditional foods of humans. However, while a diet rich in wholefoods wins hands down over a conventional Western diet (one that's high in sugar, salt, carbohydrates and meat), it's not the complete answer where the environment is at stake.

Many wholefood advocates promote an overly large consumption of meat and some of the 'trendiest' food products are not sustainable in the true sense of the word, especially when they travel great distances across the world to reach our kitchens. Two examples (for the Australian kitchen) that fail on the food miles criteria alone include Himalayan rock salt and Canadian maple syrup. More sustainable choices would be an Australian mineral salt (such as salt flakes sourced from the Murray Darling Basin) and a local, raw honey.

While wholefoods tick many of the sustainable eating boxes, in order to have a greater impact on our health and the health of our planet, we need to venture beyond wholefoods and embrace ecotarianism.

How to become an ecotarian

There are many sustainable ways of eating, including wholefoods, vegetarianism and veganism. But with the majority of the world's population simply not ready or willing to completely end their love affair with meat and animal products, not to mention many of the meat substitute products being highly processed and packaged, a different message is required to successfully drive change.

It's time for us to embrace ecotarianism.

Eat a diet rich in wholefoods, locally sourced and organic where possible.

Being ecotarian is all about eating consciously and includes:

- » avoiding packaged foods and drinks (yes, drinks are part of a diet too!)
- » reducing your meat intake and increasing consumption of plant-based foods
- » eating a diet rich in wholefoods, locally sourced and organic where possible
- » eating foods that are grown and produced locally (including foods you have grown yourself or supporting local farmers' markets)
- » eating fresh foods that are in season (rather than eating foods that have travelled great distances from other climates)
- » preserving, fermenting or drying fresh foods to enjoy when out of season
- » being super mindful of meal planning and loving your leftovers in order to eliminate food waste.

Let's take a closer look at how you can achieve three of the main ecotarian pillars; reducing processed and packaged foods, reducing meat consumption and increasing your consumption of wholefoods.

1. Reduce consumption of highly processed and packaged foods

If you shop in a regular supermarket, stick to the perimeter of the store and out of the aisles. The perimeter typically contains fresh food such as fruit, vegetables and meat. The aisles tend to contain highly processed foods like biscuits, lollies, chips and snack foods that have essentially been made in a laboratory and are high in preservatives.

Be sure to pack your reusable produce bags and purchase loose produce to reduce unnecessary packaging waste. Farmers' markets are a great place where you can find fresh and seasonal produce as well. Often the prices are more affordable because you're buying directly from the farmer or local producer.

Now, I know you may feel that you don't have time to cook or learn how to cook differently, especially if you're currently cooking with packet mixes and jars of casserole sauces. You may also think that increasing the portion of wholefoods in your diet is going to require time you just don't have. If that's you, my suggestion is to cook in bulk. If a meal is worth making, it's worth doubling or tripling. One serve is for dinner that night, the other can be enjoyed two nights later or go into the freezer for another time.

Investing in a chest freezer will allow you to cook and freeze in big batches and rotate meals to add variety in your weekly meals, without cooking seven different meals each week. While I'm our family cook, I don't cook every single night.

I generally cook one or two big meals per week and rotate previously cooked meals from the freezer. This provides us with a variety of meals throughout the week and makes more efficient use of my time.

In addition to cooking in bulk, spending a small amount of time before your grocery shop each week planning your weekly meals can save hours in the kitchen, improve your nutrition and reduce your environmental impact.

2. Reduce meat consumption

Reducing your meat consumption requires you to get more savvy with how you incorporate meat into your cooking. I'm not saying you need to give up meat in entirety (although that would significantly reduce your environmental footprint). Just start by reducing your meat consumption and opt for quality over quantity, choosing grass-fed, organic and ethically sourced meats. There are some great plant-based meat substitutes on the market these days, but be sure to read the fine print. They're not a sustainable choice if they're full of preservatives and additives, are wrapped in plastic, have been refrigerated for months and travelled thousands of food miles to arrive on your plate.

Switching out at least one meat-based meal a week for a plant-based alternative is a good place to start—make Mondays meat-free in your home! If you do choose to eat meat, opt for quality, ethically produced organic or grass-fed varieties, and stretch the meat as far as you can by making meat the bridesmaid, not the bride. So rather than having a steak and

Organic foods are better for your health and the health of our environment.

three veg for dinner, place that same portion of steak into a stir-fry with loads of vegetables and a side serve of rice or noodles. You can then transform one piece of steak that would have served one person into a meal that easily serves a family of four or five. If you're chasing protein just add some nuts or seeds to the top of your stir-fry to serve.

3. Increase consumption of local, in-season and organic wholefoods

It's quite depressing seeing organic carrots priced at $6 per kilogram, compared to $2 per kilogram for conventional carrots. I often joke that retailers should be charging more for the dose of chemicals on conventional produce and discounting the chemical-free, organic varieties. But sadly, organic production methods are typically more labour intensive and that's reflected in the higher cost.

Organic foods are better for your health and the health of our environment. Reviews of multiple studies show that organic varieties provide significantly greater levels of vitamin C, iron, magnesium and phosphorus than non-organic varieties of the same foods. While being higher in these nutrients, they are also significantly lower in nitrates and pesticides.[34]

Purchasing organic food can be quite a considerable cash outlay but there's no rule that says you need to eat all organic. Rather, get smarter about the organic foods that you do buy as not all foods have the same amount of toxins. Some fruit and vegetables are produced in a much more chemically intensive way as these crops may be more susceptible to pests.

Each year the Environmental Working Group releases their *Shopper's Guide to Pesticides in Produce*. This guide includes the Dirty Dozen and Clean Fifteen lists. The Dirty Dozen lists produce with the highest residue concentration of pesticides and herbicides and the Clean Fifteen starts at the other end of the scale, listing the 15 fruit and vegetables with the lowest concentrations of residues. Topping the 2022 Dirty Dozen were strawberries, spinach and kale, while avocados, sweet corn and pineapple topped the 2022 Clean Fifteen list.[35] If finances are tight, use this resource to inform your decision to switch some of your staple foods to organic and you'll reap the health benefits.

You may still be thinking that you don't have the budget for organic food. This is where food co-ops come in. You can join or start a group in your local community that sources bulk organic wholefoods to get great discounts. Many plant-based foods like legumes and rice are also much more economical than animal products. If you start to switch out a couple of meat-based meals for plant-based meals, you will come out on top financially.

Final thoughts

Eating sustainably is quite possibly the biggest impact we can make daily to improve the condition of the global environment. Choosing to eat organic, local produce, taking steps to avoid processed foods and reducing your meat consumption will make a significantly positive change to your health, bank balance and the state of the world.

Take action now!

It's over to you to take the steps required to eat more sustainably. You can start by taking the following three actions:

» Break up with one processed food in your diet. What can you give up today? What will you switch it out for? You can take small steps along the way! When I initially broke up with diet cola, I switched to mineral water so I could still have the bubble effect, and I drank that for a little while. But of course, mineral water has an environmental impact and a drink mile impact as well. Nowadays I happily drink filtered rainwater and herbal tea. But the jump wasn't straight to filtered rainwater from diet cola and your switch can involve a transition product too.

» Replace one weekly meat meal with a plant-based meal. You can look to increase it to two or even three later, but for now, take the initial step to one.

Use the Dirty Dozen list to help you prioritise what to switch or buy organic pantry staples like rice or legumes in bulk.

» Finally, switch one food staple from conventional to organic. Use the Dirty Dozen list to help you prioritise what to switch or buy organic pantry staples like rice or legumes in bulk. You'll be improving your health as well as the health of our environment.

Share what changes you've made to become ecotarian on social media with the hashtag #sustainabilityinthesuburbs and tag me @lauratrottadotcom so I can personally congratulate and thank you.

CHAPTER 4
Just grow something

"To plant a garden is to believe in tomorrow."
—Audrey Hepburn

My late father-in-law Marco Trotta was an exceptional vegetable gardener. He grew up in southern Italy and migrated to Australia in the early 1950s as a young man motivated to create a better life for himself and his future family. Throughout his childhood in Italy, his large family had cultivated an extensive home garden of fruits and vegetables, trading produce with others in the community for items like salamis and pasta. They were self-sufficient, like most families living in post-World War II southern Italy.

Marco continued to nurture a backyard vegetable garden throughout his life. In his early seventies, Marco and his wife, (my mother-in-law, Maria) downsized from a house to a small unit. They purchased a block of land a short walk away from their home in suburban Adelaide where Marco planted and maintained a very productive garden. Marco continued to keep his family in supply of many fruits and vegetables including apricots, tomatoes, capsicum, zucchini, cucumber, broad beans, green beans, rappa, olives, lettuce and garlic. Marco's garden gave him immense pride and happiness, gave him purpose in his twilight years and kept him physically and mentally strong.

In 2007, Paul and I married and purchased our home in the remote town of Roxby Downs. Marco and Maria visited us only a few weeks later as Marco was keen to help us install our new garden. Together, we worked for an entire weekend erecting the shade structures we'd need to protect what we planted from the harsh desert climate. We also planned what to plant and when. The knowledge and passion he shared with us during this time was invaluable.

Marco and Paul erecting a shade cover in our garden

In this chapter I seek to share the benefits of vegetable gardening with you. I will share everything Paul and I learnt from Marco about the steps you can take to ensure a successful harvest, regardless of your soil type or location.

Benefits of vegetable gardening

Growing your own vegetables and fruits has so many benefits. Whether it's herbs for your pasta sauce, tomatoes for your salad or potatoes for your roast, you simply can't go past the flavour and nutrition of homegrown produce. With the cost of living continuing to rise, and climate change and supply chain disruptions impacting fresh food supplies, growing your own vegetables also makes financial and logistical sense. Feeding your family and yourself throughout the year with produce you've grown and preserved provides a sense of self-sufficiency, freedom and pride like no other.

Growing your own food also improves your nutrient intake. English spinach, for example, retains only 53% of its folate and 54% of its keratin after eight days of being stored at refrigerator temperatures. English spinach is such an easy plant to grow. You can easily maximise your nutrient intake by growing it yourself and eating it straight out of your garden!

Growing your own fruits and vegetables also reduces your food miles. The produce we buy in supermarkets has often travelled great distances, sometimes thousands of kilometres, emitting large amounts of greenhouse gases in refrigeration and transport to make it to your kitchen. By contrast, there

There are no food miles when ingredients for your salad have travelled just metres to your plate.

are no food miles when ingredients for your salad have travelled just metres to your plate.

Many fruits and vegetables sold in your local supermarket have been treated with chemicals to make them easier to store and transport over long distances. According to Choice, a chemical called 1-methylcyclopropene (1-MCP) is used to block some of the biochemical changes that occur as fruit ripens and matures. Unfortunately, tests show the fruit remains hard and 1-MCP also prevents production of chemical compounds that contribute to flavour. Nonetheless, it's already used extensively for apples, and increasingly for other fruit, such as avocados and melons. Several fruits are also often sprayed with fungicides to prevent mould while imported fruits and vegetables may be fumigated with methyl bromide to comply with quarantine regulations.[36]

The benefits of gardening are well-documented and range from improving our mood and lowering cortisol levels (and, therefore, stress), to more significant health benefits such as reducing stroke and heart attack risk by up to 30% and dementia incidence by 36%! Luckily gardening isn't just great for adults; children benefit greatly too! From establishing strong connections with nature at a young age, encouraging healthy eating habits and providing a relaxing outlet away from the stimulation of the modern world, children who garden are healthier and happier than those who don't.

These benefits alone are enough to make many of you head to your garden but wait, there's more! Gardening has countless health and wellbeing benefits for adults and children alike, including:

1. Gardening is great exercise

All the lifting, hauling, shovelling, bending and stretching that accompanies a session in the garden can put the best circuit class to shame when it comes to a total body workout.

2. Gardening helps us produce vitamin D

Spending time outdoors with some of your skin exposed to sunlight (without sunburn) encourages the body to produce vitamin D, which is essential for a healthy immune system and strong bones and muscles. Low vitamin D levels are associated with illnesses such as Type 2 diabetes, heart disease, multiple sclerosis and some forms of cancer.

3. Gardening maintains hand strength

As we age, diminishing dexterity and strength in the hands can gradually narrow the range of activities that are possible or pleasurable. Gardening maintains dexterity and strength in the hands as we age without the need for exercises such as a physiotherapist might prescribe.

4. Gardening reduces risk of disease

An extensive Stockholm study showed that regular gardening cuts stroke and heart attack risk by up to 30% for those over 60 years. Daily gardening has also been found to represent the single biggest risk reduction for dementia, reducing incidence by at least 36%.[37] Gardening utilises so many of our critical functions, including strength, endurance, dexterity, learning, problem solving and sensory awareness. Its benefits are likely to be far-reaching when it comes to our health and prevention of disease.

5. Gardening improves your immune system

The 'friendly' soil bacteria *mycobacterium vaccae*—common in garden dirt and absorbed by inhalation or ingestion on vegetables—has been found to alleviate symptoms of psoriasis, allergies and asthma: all of which may stem from a compromised immune system.

6. Gardening improves mindfulness

Gardening requires you to take notice of your environment, to observe the weather, patterns and colours, and the cycle of life. This is an exercise in mindfulness, of being present in the moment. There is growing evidence of the positive wellness benefits of mindfulness practice, especially where good mental health is concerned.

7. Gardening improves your mental health

Gardening helps to prevent and alleviate depression and anxiety. The benefits appear to spring from a combination of physical activity, awareness of natural surroundings, cognitive stimulation and the satisfaction of the work.

Swapping your homegrown produce with other gardeners in your neighbourhood is a great way to connect with others.

8. Gardening connects people

Whether it's swapping tips, cuttings or produce with your neighbours or connecting with others in your local community garden, gardening brings people together and encourages community spirit. The street food movement encourages growing, sourcing and eating fresh food in the public space. Swapping your homegrown produce with other gardeners in your neighbourhood is also a great way to connect with others and enjoy a wider range of nutrient-dense fresh, seasonal produce.

You can even do this within your wider family. My sister-in-law Trish gifts me buckets of fresh apricots and nectarines from her trees every Christmas. My Boxing Day is typically spent watching the Test cricket on TV while preserving apricots and making nectarine chutney to enjoy all year around. In return, Trish receives some preserved apricots and nectarine chutney, along with tomatoes, cucumbers, eggplant and warrigal greens from our garden.

How to grow your own vegetables

Now that I've convinced you of the incredible benefits of growing your own food in your very own garden, let's look at how to do it effectively. I'm going to share some essential things you need to know when starting your own vegetable patch, plus some traps to avoid.

1. Choose your location and watering method

Choosing the perfect location for your vegetable patch or pots will influence just how successful your garden will become. In the southern hemisphere, a north to northeasterly aspect is favoured for growing vegetables since this position receives the most sun for the longest period throughout the year. In the northern hemisphere, a south to southwest aspect is favoured.

The cooler your climate, the more critical it is that gardens receive sunlight throughout the day. In warmer climates, sunlight is still important, but vegetables can become very heat stressed by midday and tend to grow better when shaded. Our backyard vegetable garden in Roxby Downs, one of the hottest towns in Australia, needed thick shade covers for much of the year to prevent our produce, especially the tomatoes, from literally cooking on the bush.

Along with choosing the location for your garden, you'll also need to give thought to how you're going to water your plants. If you don't want to be out hand watering your plants every day, it pays to install an irrigation system in your vegetable patch. Drip irrigation systems are ideal as they're easy to install, are cost-effective, and they minimise water loss through evaporation. They can also be tailored to provide different volumes of water for different sections of your garden, depending on the water needs for your crops. When watering your garden or planning for your irrigation system, be sure to water during the early morning (winter) or in the evening (summer) when evaporation rates are low. Avoid wetting the foliage of plants to reduce the incidence of some fungal diseases and soak the soil well, rather than sprinkling the surface. Avoid using greywater on vegetables. Instead, divert that water to shrubs and ornamentals.

2. Prepare your soil

Preparing your soil prior to planting is almost as important as choosing the ideal location for your vegetable patch. This is a gardening lesson I learnt the hard way. Christopher, our second son, wasn't the greatest sleeper as a baby and my husband Paul and I grew tireder and tireder as the sleepless nights stretched over months and years. Not long after Christopher was born, Paul, the greener thumb in our marriage, announced that he was taking a year off gardening because he just didn't have the energy or the time. He announced this in early spring when the weather was perfect for growing. I had a toddler at home to keep entertained, as well as a baby, and I really wanted the fresh organic vegetables from our vegetable garden. I took matters into my own hands and over a few days when Christopher napped, two-year old Matthew and I planted out our whole garden.

But I didn't prepare the soil first, so not one plant grew.

I subsequently learnt that our soils were very alkaline. If I had tested them with a pH meter at that time, they would probably have recorded a reading over pH 10. For vegetables

Before you start to grow your own vegetables, herbs and fruits, you need to confirm your soil is healthy.

to grow successfully our soil needed substantial preparation and the addition of compost prior to planting seeds or seedlings.

Healthy soil is the basis of a successful organic crop. The better quality your soil, the more nutrients there'll be in your produce. Before you start to grow your own vegetables, herbs and fruits, you need to confirm your soil is healthy. You can avoid a failed crop by adding nutrients to your soil before you plant and keeping an eye on your soil's pH level. Incorporating plenty of organic matter (i.e. compost) into the soil not only improves its structure, texture and nutrient content, it also provides bulk to increase the volume of the soil. This helps to improve drainage and increase the depth of soil available to grow vegetables.

It pays to add compost to your soil at the start of every planting season. There are loads of options for compost or fertiliser to add to your soil, but animal manure is our preferred compost. Adding a good dose of animal manure to your vegetable garden gives it a vital nutrient boost. Animal manure also increases microbial activity in the soil and improves drainage and moisture retention in sandy soils. Our favourite animal manure to use in our garden is chicken manure, mostly because we have backyard chickens. But the tip here is not to use fresh chicken manure because it's hot and will burn your plants if you're applying it directly. Alternatively, invest in a garden compost bin for your chicken manure so it can break down over time before adding to your vegetable garden a few months later.

There are pros and cons of different animal manures and soil additives. At the end of the day, what you can source locally or produce in your own backyard is often the best and easiest option. When adding compost to the soil, be generous and use one full wheelbarrow per square metre.

Once you've prepared your soil by adding some compost, there's one crucial soil characteristic you need to test before planting your seeds, and that's pH. Soil pH is a concentration of hydrogen ions in the soil and is an indication of the acidity or alkalinity of soil. pH directly affects the concentration and availability of major nutrients, the forms of microelements available per plant uptake, as well as biological activity in the soil. Bacteria, for example, are less available in acidic soils.

The optimal pH range for most vegetable species is between pH 5.5 and 8. Outside this range, most plants will suffer nutrient deficiencies and toxicities. However, like everything in life, there are exceptions. The saltbush that grows throughout Outback Australia thrives in a highly alkaline soil, whereas vegetables do not.

The natural tendency for most soils is to become more acidic over time. To counter this and increase the pH, limestone primarily made from calcium carbonate or dolomite, which is magnesium carbonate, can be added to the soil. Garden lime is readily available at hardware stores and garden centres. Once blended in, test the pH again and adjust if necessary.

There are a few options for treating alkaline soils. The addition of a good quality compost helps, as does adding blood and bone fertiliser, or an acidifying agent such as sulfuric acid, elemental sulphur or pyrites. If you have a large vegetable garden, the cost to treat alkaline soil can add up. The addition of gypsum can reduce alkaline pH to some extent. Growing legumes in crop rotation may also help in sustaining any pH reduction.

We struggled with alkaline soil for a long time in Roxby Downs. What worked for us was a combination of chicken manure, commercial compost, with the addition of sulphur (just into the immediate planting area only, not the entire garden, to help keep costs down).

3. Plant the right plant at the right time

You've chosen your location, set up how you're going to water your plants and prepared your soil. It's time to get planting but what should you plant and when?

Growing your own vegetables from seed is very rewarding and has several advantages over seedlings. Seeds are inexpensive and, if you harvest your own seeds, subsequent crops are free. Growing plants from seeds also allows you to access a greater number of varieties, such as heirloom vegetable varieties. Some vegetables, such as peas, transfer poorly and are best planted as seeds directly where they'll be harvested. Most vegetables grow very easily from seed, and you can play an important role in collecting and saving seeds from heirloom varieties if you plant and collect the seeds from them.

You may want to grow something but don't have much space. Pots are the answer here.

Despite these advantages, if growing vegetables from seed is just one extra barrier between you and gardening, and you're more comfortable starting with seedlings, or even cuttings from a friend's garden, then go for that option. It's much better to grow something than nothing. You can always switch to seeds down the track as you gain more confidence in the garden.

Before you dash off to the shops to buy some seeds or seedlings, you'll need to decide what to plant. Where you live in the world and your local climatic factors have a significant impact on what vegetables to plant and when. Download my planting guides at **plantingguide.lauratrotta.com** for more information on the best season to plant for your climate.

Don't wait, just start

I know so many people who love the idea of growing their own fresh produce, but just *don't know where to start*. If this is you, start by planting just one thing. Maybe you buy baby spinach leaves every week, or coriander, or cherry tomatoes. Plant something that:

>> you eat regularly and
>> will grow well in your climate.

You'll be motivated to grow it and keep it alive because it's one of your favourite vegetables or herbs to eat.

You may also want to grow something but *don't have much space*. Pots are the answer here. In my early twenties, I lived in an apartment in Townsville, North Queensland. I was sick of buying fresh herbs wrapped in plastic and wanted the economical option and flavour hit from extra fresh herbs. So I started growing herbs in pots on my windowsill. There are also many clever vertical garden configurations available these days for apartment dwellers. Have a look at what's available and see if something suits your living arrangements.

Another reason people who wish to grow their own food don't start is because they believe they *don't have the time*. It's worth remembering here that we will always make the time for things we value and benefit from. Growing your own fruits or vegetables is something you'll never regret prioritising.

The final barrier to starting for many people is because they feel they *don't have a green thumb*. It might help you to know I've had many experiences where I've killed everything I planted. But it always came down to one of:

» not preparing the soil correctly (or not preparing it at all!)

» neglecting my plants

» not watering them adequately.

If you choose the right location for your garden, prepare your soil, plant the right varieties for your climate at the correct time of year, and set up an adequate watering system for your plants, your garden will succeed and you'll be eating fabulous, organic, nutrient-dense, zero food mile produce daily.

Final thoughts

Growing your own food, even just a small amount, lowers your environmental footprint, improves your health and wellbeing and saves you money. If you haven't yet dabbled with growing your own vegetables, make the effort to start and you'll be rewarded considerably for your time and efforts.

Take action now!

It's over to you to start or expand your vegetable garden by planting ONE thing. Complete the following three actions to maximise your gardening success:

1. Choose the vegetable, herb or fruit you will plant.
2. Select your location and decide how you will water your plant. If planting directly in your soil, test the soil pH using a kit from your local hardware or gardening centre and add compost or sulphur to ensure your soil is as neutral pH as possible.
3. Plant your seed or seedling/s.

Share photos on social media with the hashtag #sustainabilityinthesuburbs and tag me at @lauratrottadotcom so I can congratulate you!

CHAPTER 5
Work towards a zero-waste home

"We don't need a handful of people doing zero waste perfectly. We need millions of people doing it imperfectly."
—ANNE-MARIE BONNEAU

Do you have those chores in your household that everyone hates doing? You know the ones—they kind of just sit there undone, looking messier and messier until someone gives in and does the job out of frustration?

In most houses I've lived, especially share houses, that job was emptying the rubbish bin.

This reached its most comical levels when I lived with my elder sister Kate while we were both studying full time at university. Chores got evenly divided up to make sure that everything was fair, but we hit a stalemate over taking out the rubbish. Neither of us wanted to do that one.

And so the game began …

We were both ninjas at squeezing just another bit of waste into the bin without it overflowing. Until the inevitable happened and you just physically couldn't fit another piece in. If it was your piece that tipped the bin over the edge, it was up to you to take the rubbish out.

Fast-forward a few decades and I still dislike taking out the rubbish. But instead of being stubborn and jamming the bin to bursting point, I've taken a different approach, and that's to reduce the amount of waste we generate in the first place.

I'm so thrilled to say that I hardly ever need to take the rubbish out because (pause for dramatic effect …) most weeks we just don't produce any.

That's right, we're a family of four that barely produces one tiny bag of waste a week.

Recycling, yes, but waste, no.

I won't say getting to this point happened overnight. Rather, a series of small changes made over time has basically meant I retired long ago from taking out the rubbish.

And before you tell yourself that this can't happen for your household, I want you to instead ask yourself why this couldn't happen and how it could happen for your household … because there's absolutely no reason why it can't happen for you.

What is a zero-waste household?

A zero-waste household avoids waste being the end result of consumption and strives instead to be circular (there's infinite purpose in the products we create and use).

I realise the terms 'waste-free' home and 'zero-waste' home may seem farfetched for you. I want to stress that these terms are aspirational and how far you go in reducing your household's waste is totally up to you. Like everything with sustainable living though, you'll find that by reducing your household's waste:

» You'll feel a lightness and freedom you may not have felt in years.
» Your home will be a healthier place to live.
» You'll save a load of cash on the regular purchase of single-use items!

But the best benefits of a zero-waste home are convenience and freedom.

For years, single-use items like nappies, tissues and serviettes have been marketed as time-saving items of convenience. But panic buying and supply chain issues prevalent during the COVID-19 pandemic highlighted to households that running out of these items was indeed very *inconvenient*. By contrast, families using reusable items like cloth nappies, wipes, handkerchiefs and napkins were unimpacted by the shortages of their single-use alternatives.

*Reducing our household
waste is necessary because
the waste we produce
in our home is literally
suffocating us and
our planet.*

In this chapter I'll share some of my best tips for creating a zero-waste home. I'll focus in particular on how you can easily reduce the two biggest household waste streams: food and plastic.

Why zero-waste?

Australians produce 540kg of household waste per person each year, which is more than 10kg for every single person, every single week.[38]

Reducing our household waste is necessary because the waste we produce in our home is literally suffocating us and our planet:

» We're burying items in landfill that will take hundreds of years, (if ever) to break down. Plastic alone takes over 500 years to break down. This means that every disposable nappy, sanitary pad, plastic straw, plastic film wrap and supermarket collectable toy ever tossed aside is still laying in landfill, intact.

» Of the organic wastes that do decompose, most release greenhouse gases like methane in the process. This contributes significantly to climate change.

» Our oceans are becoming a toxic soup of microplastics that are accumulating in the food chain and impacting marine and birdlife in the process. It's estimated that about 130,000 tonnes of Australian plastic ends up in our waterways and oceans each year.

While the increase in kerbside recycling in recent decades is helping, we still have a long way to go. Of the estimated 67 million tonnes of waste Australians generated in 2017, just 37 million tonnes were recycled.[39]

The waste we generate is changing our landscape, changing our oceans and changing our atmosphere. It's changing our environment and because we're part of our environment, it's changing us. Waste is a serious environmental issue and at first glance it can appear too hard to solve, but it's not.

Let's look at what you can do to reduce your household waste by:

» exploring the concept of a circular rather than linear waste household

» unpacking the six principles of a zero-waste home

» looking at 12 ways you can reduce food waste.

Moving from a linear to a circular household

Consumption of resources in most households follows a linear model. That is, waste is a function of items brought into a home, minus items consumed and stored.

Figure 2: Linear household model

This model is inherently flawed as it's based on the assumption that Earth has an infinite amount of resources as well as infinite regenerative capacity.

Under this model, to reduce household waste we'd cut back on the items we're bringing into our home and reduce our consumption to ensure we're minimising the amount of unwanted or waste items we're tossing away. But in a world where the human population and consumption habits continue to grow exponentially, the mere act of reducing our household waste is not enough. We need to move beyond the linear model and embrace a circular model, where any waste we create is a resource for another process, either within or outside our home. I share practical examples of how to do this in the next two sections.

Rather than reducing the harm we're having on the environment, we should be actively seeking to avoid it in the first place.

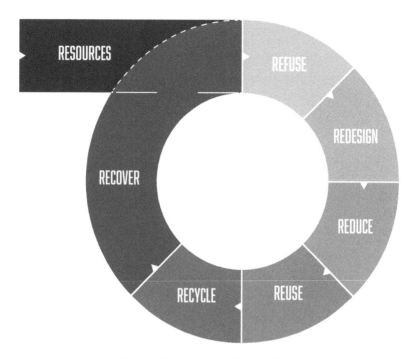

Figure 3: Circular household model

Six principles of a zero-waste home

In 1976, the US Congress passed the *Resource Conservation and Recovery Act* to increase recycling and conservation efforts as waste became a bigger problem. The slogan 'reduce, reuse, recycle' was born at this time.[40] While this mantra has inspired society to be more conscientious of our waste and the planet's health, there are new elements to be added in to bring it in line with modern thinking and capabilities.

The first thing we need to acknowledge is that rather than *reducing* the harm we're having on the environment, we should be actively seeking to *avoid* it in the first place.

How? By refusing to buy into excess and consumerism. By avoiding the purchase or consumption in the first place, we're saving resources both upfront (creation) and at the end (disposal or reprocessing) of the product lifecycle.

Furthermore, the mantra currently stops at 'Recycle' but technology nowadays enables us to reprocess or recover waste even further to produce a valuable resource. In practice this can be the capture of methane gas from landfill for power generation or as simple as composting your food scraps at home to produce fertiliser for your garden.

Let's take a closer look at how we can expand the phrase 'Reduce, Reuse, Recycle' to illustrate the principles of a zero-waste home.

1. REFUSE

As a high-end cosmetic fan in my early twenties, I collected a mountain of travel size, luxury lipsticks, mascaras, eye shadows and creams via 'gift with purchase' promotions. These items were of little use when I was working in the field in my high visibility overalls, steel-capped boots and hard hat as an environmental engineer. Unsurprisingly, after sitting idle in a drawer in my dressing table for a decade or two, I finally threw the cosmetics out in a house declutter. They were so long past their expiry date that they weren't even suitable to give away. For this I feel terrible but I learnt my lesson and these days no longer fall for the 'gift with purchase' hook. (In fact, I've long since given up fancy cosmetics, but that's a story for another chapter.)

This first principle of a zero-waste home simply encourages you to avoid the purchase or consumption in the first place:

» Do you really NEED the latest gadgets and gizmos?

» Does your clothing always need to reflect the latest trend?

This principle extends to smaller items as well. Do you need:

» A straw with your drink?

» A plastic bag with your purchase?

» The disposable cutlery that comes with your takeaway?

» Small shampoos and conditioners from hotel rooms?

Most likely, you don't.

Instead, become a conscious consumer and question every purchase or acquisition, and avoid single-use items. By only taking what you truly need (with the occasional want thrown in there to stop you feeling deprived) you'll be stopping waste where it counts, at the source.

2. REDESIGN

Sustainability and music are my two main passions, and I started my music collection from a young age. By the time I was ten years old I was regularly spending my pocket money on music cassettes, and then CDs. By my mid-twenties my CD collection needed a large storage unit, carry cases to take

One question you can regularly ask yourself when it comes to items in your home is "is ownership really necessary?"

with me on my travels, not to mention the sound system and Discman to play my favourite tunes. My music collection brought me much joy but there was undeniably a waste component to it.

Technology quickly advanced in the 2000s, leading me to donate most of my CD collection to a community radio station in our declutter prior to our last house move. I now pay a monthly subscription to stream any song or album I wish, ad free, and I'm happy with that arrangement. Smartphones and music streaming services together have eliminated the need for CDs, Discmans and CD players. It's just one example of how processes can be redesigned to eliminate waste.

Designing out waste is key to a circular household, yet the reality is that most items today are still designed for the linear model. One question you can regularly ask yourself when it comes to items in your home is "is ownership really necessary?" Once you embrace renting, streaming or borrowing and returning items when you've finished with them, you significantly reduce the number of items you need to manage and ultimately the volume of waste from your home.

3. REDUCE

The sheer fact that we're living and walking on this earth means that we'll have an impact, but there are many ways in which we can minimise the impact we have.

Simple actions you can take to reduce the waste you generate include:

- » buying in bulk to reduce packaging waste
- » choosing products that have less packaging (or compostable packaging)
- » opting for quality purchases that will stand the test of time rather than buying a larger number of cheaper goods
- » eating the food you purchase to reduce the volume of food wasted.

Another strategy is to simply wait for longer periods before you replace an item. Can you put up with your perfectly functional smartphone for another year before upgrading to the latest model?

A quick note on packaging and greenwashing around the term 'biodegradable': choose *compostable* packaging over *biodegradable* packaging. The former is organic and breaks down into soil relatively quickly at the end of its lifespan, whereas the latter just breaks down into smaller pieces of plastic that can enter the food chain.

4. REUSE

This step involves the adoption of the mindset of 'single use sucks'. Bit by bit, replace as many single-use items in your home and workplace as you can with reusable alternatives. Some of my favourite reusable items include:

- » Reusable food covers or containers instead of covering foods in plastic wrap
- » A menstrual cup rather than tampons

» Handkerchiefs rather than tissues.

As a herbal tea drinker, I carry a thermal flask of hot water with me when out and about so I can enjoy a cup of tea wherever I please, without the waste of a disposable cup from a café.

5. RECYCLE

A few years back when my husband and I were designing the kitchen in our current home, I requested separate receptacles for different waste streams tucked inside the kitchen cupboards. The kitchen designer tried to talk me out of it several times, saying I was wasting valuable cupboard space. But I stood firm as I wanted a separate compartment for recyclables, one for soft plastics, one for food scraps and one for general waste, all out of sight behind a cupboard door. The alternative was what we had in our previous home—a big tub for recyclables on the floor in the middle of the kitchen, beside the refrigerator and a bag for recyclable soft plastics right next to it.

Functional? Yes. Attractive? Absolutely not.

The kitchen designer reluctantly agreed to do some research and found German-designed cupboard drawers for waste and recyclables. I ended up getting my functional and stylish kitchen waste segregation system and love its ease and efficiency. Regardless of how functional or stylish your kitchen waste segregation system is, the principle is the same, to make it easy for your household to recycle whatever you can.

As technology continues to advance, there will be more and more opportunity to recover energy from waste.

It's important to remember, however, that recycling also uses resources. It's better to avoid the generation of waste in the first place than use energy and chemicals to recycle items. One area we particularly need to improve is e-waste recycling.

E-waste is any electrical or electronic item that needs a plug or battery to work. With ever-advancing changes in technology, households can have numerous electronic gadgets, many of which only last a few years. E-waste is the fastest growing waste stream in Australia and contains many toxic substances such as lead, mercury, cadmium and lithium. In landfill, e-waste leaches these hazardous substances into the soil, groundwater and surface water which can lead to health and environmental problems. Global e-waste is expected to hit over 70 million tonnes per year by 2030 and despite the waste stream containing gold and silver, valuable glass and rare earth elements, only about 17.4% will be effectively recycled, based on 2019 figures.[41]

6. RECOVER

As we get smarter with our waste management systems and technology continues to advance, there will be more and more opportunity to recover energy from waste. Many local councils are already capturing gases from landfills to use as energy sources and recovering by-products of sewage for fertilising degraded lands. We can employ this principle of zero waste on a smaller scale in our homes too!

What waste stream are you producing in your own home that can become an energy source for another purpose? Food waste is a great example here; by either composting your food scraps or feeding them to animals like backyard chickens or a worm farm, you're turning your waste into a valuable resource such as organic compost for your vegetable garden or ethical eggs.

Now that you understand the six principles of a zero-waste home it's time to take a closer look at one of the main waste streams leaving our homes, food.

How to reduce food waste

Like many children of the eighties, I was made to eat everything on my dinner plate.

A typical evening meal in our household was accompanied by TV commentary of the terrible famine gripping Africa and my father drumming into us how fortunate we were to live in the 'Lucky Country'. It's not surprising then, that more than three decades later I'm continually looking for ways to reduce my family's food waste and feel guilty on the odd occasions when I do throw out food.

It's fair to say that food waste is a by-product of our affluent and time-poor society.

In Australia, food is the single largest component of household waste. Australian households throw away one in five bags of groceries, equal to around 312 kilograms per person per year.[42] Can you imagine going to the supermarket and buying five bags of food, only to leave one behind on the counter? That's what it's like week in week out in the average Australian home as much of the food that's tossed is edible.

Every time we throw food in the rubbish bin, we're not just wasting our money. We're discarding the vast amounts of resources, energy and water that it took to produce, process, store, refrigerate, transport and cook the food. Plus, rotting food in landfill gives off methane, which, as you know, is a greenhouse gas that is particularly damaging to our environment. Luckily, reducing our food waste isn't as hard as you might think. Here are 12 small steps you can take to reduce your food waste.

1. Plan your meals

Meal planning is the critical ingredient in reducing your food waste. I find that by planning each meal for the week ahead, I'm less likely to buy too much food in the first place. Include plans to eat out and leftover meals in your meal plan to avoid purchasing fresh food you won't end up cooking before it spoils. Once you have your meal plan sorted, create your shopping list directly from your plan, taking into account the food stocks you have on hand.

2. Check your pantry, fridge and freezer contents before you shop

Too often we make spontaneous purchases when shopping only to bring the produce home and discover that we already

When unpacking groceries, move older products to the front of your pantry, fridge and freezer.

had adequate supplies. To avoid doubling up, check stocks of food in your pantry, fridge and freezer BEFORE you shop and remove any item from your shopping list that you already have.

3. Shop to your plan

Of course, a plan is only as good as the execution so when shopping, stick to your list. If your meal plan tells you that you need three potatoes, don't buy five kilograms of them. Of course, it makes financial sense to stock up on specials of pantry staples you use often but be extra careful where fresh produce and perishables are concerned. You won't have saved anything if you end up tossing the produce a few days later. If you're particularly swayed by specials and marketing tactics, shop for produce online and try to shop from your favourites list rather than browsing the entire store, to avoid impulse purchases.

4. Keep stock of your stock

When unpacking groceries, move older products to the front of your pantry, fridge and freezer and place new products towards the back to help you use up food before it expires. If you have trouble keeping track of your food stocks, develop a labelling system or place a list of contents and date to consume by in a visible location.

5. Store your food correctly

Living for several years in the tropics taught me everything I know about food storage. I learnt the hard way if I didn't store my food correctly, weevils would hatch in my flour and ants would invade anything that wasn't correctly sealed. If you're regularly dealing with pest problems or throwing away stale biscuits and cereals, I highly recommend investing in quality, airtight containers to store your food. Repurposed glass jars are an inexpensive, airtight way to store food without the expense of a designer pantry storage system.

6. Don't be overzealous with the veggie peeler

Do you really need to peel your carrots, potatoes and cucumbers? The skin on our fruit and veg is nutrient rich and full of fibre and, therefore, worth eating! I'm not expecting everyone to enjoy eating furry kiwi fruit, but just be mindful of the quantity of fruit and vegetables you discard before you even start preparing meals. If you're purchasing organic fruits and vegetables that's even more reason not to be overzealous with the veggie peeler.

7. Think before you throw

Too often we tend to throw food that, while not in its ideal form, could easily be transformed into something more appealing. Broccoli stalks are a perfect example. They add bulk to stir-fries and soups and are great at absorbing the flavour of dishes. They can also easily be frozen for when you get around to making your own broths or stocks. Overripe bananas make delicious banana bread, cakes and smoothies. Wilted veggies are perfect for stock or vegetable soup, and stale bread and lonely crusts make the best breadcrumbs for rissoles or schnitzels. Simply whiz them in your food processor, dry in a low oven or food dehydrator and store in an airtight container when cool. By repurposing foods that are past their prime into another form you're giving another lease of life to the food and avoiding food waste.

8. Learn the art of preserving

Before the days of refrigeration, preserving food was a common occurrence throughout the world to ensure adequate food supplies all year round. Preserving food is still a fantastic way to stretch your budget, reduce food waste and improve your health by eating nutrient dense foods. If you're lucky enough to receive a box of fresh produce such as apricots or tomatoes from a friend's garden, preserve them and enjoy them for months to come. If you're a green thumb, consider joining your local produce swap group on social media to trade homegrown fruits, vegetables or preserves with other gardeners in your suburb.

9. Portion control

To avoid situations where your eyes are bigger than your belly, dish up servings on a smaller plate. When eating out, split dishes with your dining companion/s to avoid tossing

By getting creative with what you do with your leftovers, you can also inject a dose of fun into your cooking.

half of the giant portions found in many restaurants or order an entrée size rather than a main meal serving. As for smorgasbords ... Don't. Even. Go. There.

10. Love your leftovers

Leftovers are not only great for meals where you're short of time, but they can also become the base for an entirely new dish. Boiled rice can easily be turned into fried rice or rice puddings, excess pasta is great in mornays or bakes, and leftover roast vegetables make a tasty 'bubble and squeak' or frittata. By getting creative with what you do with your leftovers, you can also inject a dose of fun into your cooking.

11. Use your senses

I'm sure my household isn't the only one where milkshakes are on the menu the day the milk expires. Of course, it's great to exercise caution when it comes to food safety, but in many cases, expiration dates on foods are just as much about manufacturer's recommendations for peak quality as they are for food safety. If stored correctly, most foods (including meat and dairy) will stay fresh several days past their conservative 'use by' date. If the food looks, smells and tastes okay, chances are it is safe to consume.

12. Give waste another life

Some of the foods we throw away can easily be used again. For example, I always reuse carcasses from a steamed or roasted chicken to make stock. Vegetable trimmings including onion and garlic skins, or the green section of leeks can also be added to stock to enhance the flavour. I have a large container in my freezer where I keep these vegetable trimmings until I'm ready to make a batch of broth in my slow cooker. By making your own broths and stocks from these vegetable offcuts you also save on the expense and waste associated with buying tetra pack or powdered varieties.

Despite your best efforts, there will be occasions where you need to throw out food and for these occasions I recommend backyard chickens, composting or even a worm farm. Our backyard chickens thrive on the cast offs from our meals and in return keep us in full supply of ethical eggs.

Final thoughts

Reducing household waste significantly lowers our environmental footprint and the negative legacy we'll leave on Earth. By transforming your household mass balance from linear to circular, understanding and enacting the principles of a zero-waste home, and by getting super savvy with how you purchase and consume food, you'll be well on your way to joining the growing number of people in my community who have retired from taking out their rubbish bin!

Take action now!

It's over to you to take the steps required to reduce your household waste. You can start by taking the following three actions:

1. Choose one step in the 12 steps to reduce food waste or one food item you regularly find yourself tossing out to work on. For example, if you toss out broccoli stalks, slice up and put in stir-fries or freeze and add them to your homemade stocks.

2. Take my **Single Use Sucks** pledge by choosing one single-use item you regularly purchase and switch it out for a reusable alternative.

3. Round up your stockpiles of e-waste items in your household (e.g. phone handsets, old laptops and gaming consoles) and take them to your local collection point to facilitate the recovery of rare minerals from them.

Now that we've taken steps to reduce the waste your household currently produces it's time to become a conscious consumer and reduce what you buy in the first place.

Share what change you've made on social media with the hashtags #singleusesucks and #sustainabilityinthesuburbs and tag me @lauratrottadotcom so I can personally congratulate and thank you.

CHAPTER 6
Create a minimalist home

---

*"The most environmentally friendly product
is the one you didn't buy."*
——Joshua Becker

In early 2018 my family and I relocated from the small, remote town of Roxby Downs we'd called home for 13 years, to Adelaide, a city of 1.2 million people 600 km away. In the weeks and months leading up to our move, we made a significant effort to declutter our belongings. We had a garage sale, participated in the town's car boot sales, sold a heap of items via our local buy swap sell social media page and donated carloads of items to the local opportunity shop. We thought we'd done a pretty good job at decluttering, but when the removalists arrived with the truck they use for a standard three-bedroom home, we struggled to fit all our belongings inside. Of course, there was no option to get a larger truck—the removalists had already travelled 600 km to our home to help us load. Thanks to the skill of the removalists, we managed to fit 99% of our possessions in the truck and left our laundry buckets, mops and broom behind. Yes, it was that tight.

When we made the move, it was to a new home we were building in Adelaide that was delayed by several months (so wasn't ready for us). We had no choice but to rent a short-term furnished unit and put our items in storage. I was quite surprised by the number of options for places to store our items. "Surely not everyone is moving house and needing short-term storage," I thought. "Who else has need for storage?"

It turns out there's a lot of people who have a hard time letting go of their stuff. A survey by StorageCafe in March 2022 found that 38% of Americans rent self-storage, with furniture the most common item kept in a storage unit.[43] The trend is similar in Australia where there's one self-storage unit per 11 people and where there are more self-storage units than McDonald's restaurants.[44]

In Chapter 5 we looked at reducing the waste your household is generating. Now we're going to unpack your relationship with stuff, addressing what you buy and accumulate in the first place as well as how you manage it in your home. Essentially, we're going to explore how to:

> » embrace a minimalist mindset
> » become a conscious consumer
> » live with less.

Signs your stuff is controlling you

How do you know if you have an unhealthy relationship with your belongings and clutter, (essentially your 'stuff')? Well, firstly, your credit card may be maxed out and the main reason this would usually be the case is because you're living beyond your means. You might be eating out frequently or going on holidays you can't afford, but it can also be an indication that you're accumulating stuff. Are you shopping for emotional reasons? If you are, it might be time to find a hobby or give your personal relationships some attention rather than hobby shopping.

The second sign may be that you can't park your car in your garage. Australians are buying larger houses to accommodate their growing volume of stuff, despite having the largest house size in the world (the United States isn't far behind). Many of these larger houses have garages full of belongings. If you're parking your car in your driveway or on the street because your garage is being used as a storage space, it's a sign that you need to address your relationship with stuff.

Thirdly, you may be paying for additional storage. Maybe your house and garage are full, and you have stuff in storage, or multiple storage venues. This was the case for my family when I was growing up. I can't remember a time when my father didn't have anything in storage. From a car body and car parts he was going to rebuild "one day", to certificates and training manuals from all the courses he'd completed as a police officer, to toys from my childhood as well as some

*We make internal excuses
not to part with our
stuff because we like the
memories associated
with the item.*

he'd accumulated from garage sales for future grandchildren. My father gained pleasure from his belongings and really struggled to move items on. While I don't have one memory of him tossing anything away, I did learn early in life that stuff is never a bargain if you're paying to store it.

You may also struggle to move items on and if this is you, you're not alone.

» Maybe you like remembering your first boyfriend when you look at those earrings he gave you.

» Maybe you're living in hope that one day you might fit into those size 12 skinny jeans again.

» Maybe you're holding on to a dream that in the future, you'll have a grandchild that will lie on that baby mat your firstborn used and wear the 20 newborn outfits you have stashed away.

» Maybe you're holding onto an item that someone special gave you, even though you never liked the item and have never used it.

We make internal excuses not to part with our stuff because we like the memories associated with the item or hope we might be able to use it again. But maybe, just maybe, by moving on items like these we'll be able to live in the present a bit better. I'm not saying to ditch the family heirloom, especially if you're using it. I'm just encouraging you to ask yourself why you're holding onto items you rarely or never use and be open to moving them on if they're no longer serving you.

Do you own a caravan you use once or twice per year? A boat that's on the land more than it's in the water? A trailer with flat tyres because it's been an age since you used it?

These are all big items that can easily (and comparatively cheaply) be hired rather than purchased outright. But you can also easily hire smaller items too, including lawnmowers, whipper snippers, baby items, toys and books (just think of your local library).

If you rarely rent items, that's another sign that your relationship with stuff may need addressing. Ownership is all about control and comes with a premium price tag. If you're not a fan of renting items, it might be time to ask yourself why that's the case. Once you embrace the freedom of renting you won't look back!

Do you line up for the latest iPhone or gaming device? Living for the new and needing to be one of the first to have the latest release of anything limited edition is another sign that you may be obsessed with stuff. Naturally, this hits your wallet extra hard as you're not only upgrading items well before they're worn out, you're paying a premium price to be an early adopter. If you can relate to this, I challenge you to practise patience and skip a model or two. The world will still turn!

The final sign that your stuff is starting to control you is that you avoid having visitors to your home because your spare room is full of items, or you're embarrassed that your house is so messy and cluttered. If that's the case, it's likely that your stuff is interfering with your friendships and connections. It's time to make some choices and remember that stuff will never make you as happy as healthy, connected relationships.

If you can relate to any of the points above, do not feel bad. Despite being a long-time environmentalist, I picked up some hoarding habits in my childhood that have been particularly hard to shake. It can still be an effort for me to move items on, even though I've consistently worked on my relationship with stuff over the years. But when I *do* move things on, I love the resulting freedom and space—physically and energetically. Let's now unpack some of the benefits of controlling your stuff before it controls you.

Benefits of a minimalist home

There are so many benefits when you take control of your stuff and break away from the control that it has over you. Here are six benefits of living with less:

1. Less stuff = more energy

Less clutter in your home will lead to you having more energy. This comes from having fewer items to clean and organise. But it's also a result of freeing up the physical and energetic space that clutter takes up in our homes and lives.

2. Less stuff = more money

Stuff is expensive. It's expensive to accumulate in the first place. It's also expensive to store, particularly if the volume of our belongings exceeds the storage space in our home. The sad reality for many households that use self-storage is that they'll end up paying more to store their stuff than the stored items are worth.

*We are happier and
more connected when we
accumulate less stuff.*

3. Less stuff = more meaning

We are happier and more connected when we accumulate less stuff. A study published in the *Journal of Positive Psychology* found that interpersonal relationships are a predictor of life satisfaction.[45] Stuff isn't an adequate substitute for healthy and connected relationships. It won't make you happy. Those who adopt a minimalist way of life can dedicate more time fostering relationships as they have fewer items to direct their time and energy towards. In a world where everyone is busy and distracted, minimalism allows for a deeper connection to other human beings.

4. Less stuff = less burden on others

Having less stuff also reduces the burden you're having on other people. If something happened to you and you weren't around to manage all your stuff, who would have to go through it all and manage it on your behalf? Is that something that you really want to burden other people with?

5. Less stuff = better health

Stuff is a health hazard. It can be a trip or falling hazard, but it also collects dust and harbours spiders and other critters. Reducing the amount of stuff in our homes creates a healthier and safer environment for ourselves and our families, and that alone is worth making an effort for.

6. Less stuff = better outlook

Too much stuff prevents us from moving on from our past. When you're clinging onto stuff and the memories associated with it, it's very difficult to enjoy the present and embrace the future.

How to create a minimalist home

Given our house sizes are getting larger, and the self-storage industry continues to boom, it's very obvious the majority of us are struggling to take control of our stuff. So let's start our journey towards minimalism by working through the following four steps.

1. Choose where to start

The first step is choosing where to start. Energetic space clearer Helen Joy Butler says there are five places you can start, depending on your energy levels and your ability to make decisions. The first place is the entrance to your home. It's logical to start here because it's where you welcome visitors into your home.

You can also work out from the centre of your house, likening the process to that of the wheel of a bicycle. The inside of a wheel spins much faster than the outside so you can gain momentum by starting in the middle of your house and spreading the energy from the middle to other parts of your home.

The third place is the area that causes you the most grief. Now, this isn't for the faint-hearted, so only start here if you are ready.

The fourth place is something small like the junk drawer in your kitchen or the Post-it® notes on your desk.

And the final or the fifth place to start is just start anywhere. This is perfect for those perfectionists or procrastinators who have trouble starting in the first place. If that's you, just start where you're standing and declutter and organise that space.

2. Clear the space and sort your belongings

The second step to create a minimalist home is to empty the space you're decluttering and sort everything into three piles:

- » a pile you're keeping
- » a pile you're donating
- » a pile you're discarding.

The discard pile will include damaged items or items that aren't of a sufficient quality to donate. Selling is another option, but you need to really ask yourself if an item is worth your time and energy to sell. If you do decide to sell an item, give yourself a deadline. If it hasn't sold by that time, have a Plan B like donating it to charity or giving it to a family member or friend.

*Dealing with our stuff
and deciding what to keep
and what to move on can
bring up a lot of emotions
and resistance for people.*

Dealing with our stuff and deciding what to keep and what to move on can bring up a lot of emotions and resistance for people. The most common excuse we use for holding onto stuff is that we think we're going to need it again one day. Over the past 12 years I've been four different standard clothing sizes, (more if I include pregnancy clothing in the mix) and unsurprisingly, my wardrobe reflects this variation. I found that I was keeping a large tub of size 12 clothing that I wore in my late twenties BC (before children) when I was going to the gym six days a week as well as playing regular team sport ... time I just don't have in my current life stage. I found that holding onto these items was a form of torture and once I moved the items on, I was able to accept my body much better.

If you have a large amount of clothing that no longer fits you, I advise donating the clothing to others in need and treating yourself with a quality outfit that makes you feel amazing right now. If you return to your ideal size in future, you'll likely be older, the fashions will have changed, and you may not end up wearing the items you've kept all those years anyway. So, release them and relieve yourself of the stress in the meantime.

Another challenge you may face when moving items on is that someone special gave the item to you and for that reason alone you can't bear to part with it. I can also relate here.

When Paul and I became engaged, very close family friends gifted us a cutlery set. These family friends were an elderly couple with whom I shared a deep bond of love—they treated me as if I were their granddaughter and I loved them both. Even now I still count them both as among the most important people in the world to me. Their joy in gifting this set to us was so apparent; it was a valuable gift and there was much love behind their selection. The only issue was that I'd saved up and purchased a quality cutlery set in a design I really liked a couple of years earlier and I didn't need a second set, let alone one that just wasn't my style. I accepted the gift with grace and much thanks and kept the cutlery set in a drawer, unused. I couldn't use it, but I also couldn't move it on as I'd attached so much meaning and emotion to the cutlery set. To move it on would be rejecting their love. It was only after more than a decade had passed, and my elderly family friends had passed on, that I was able to pass the cutlery set onto a family who needed and wanted the set. I was only able to take this step when I acknowledged that my friends wouldn't want me to feel burdened by their gift. I also didn't need the cutlery set to remind me of my wonderful memories of them. I had those memories and that strong bond of love inside me.

Decluttering and donating your stuff won't just make you feel amazing, it can also significantly help others. There are some fantastic charity organisations accepting donations of quality secondhand items to distribute to those in need.

After cleaning out my bra drawer I donated 31 bras in various sizes (thanks to varying across six bra sizes throughout pregnancy and breastfeeding!) to the Uplift Project. This project sends bras to women in disadvantaged communities where a bra is often unobtainable or unaffordable.

I posted four pairs of prescription glasses to Lions Recycle for Sight Australia where they were provided to people in need in developing countries.

I also donated a large box of quality jackets, shirts, pants, skirts and shoes that no longer fit me to Dress for Success. This is a worldwide organisation that distributes business clothing to women in need who are preparing for their first job interview to enter or re-enter the workforce with a job that could change their life.

3. Establish systems, routines and boundaries to prevent clutter from rebuilding

To prevent clutter, you need to:

» create systems

» have some routines in place

» establish boundaries around what's coming into your home and your schedule.

In my household we've struggled with everyone (myself included) dumping their bags when they arrive home next to the kitchen bench. This not only looks unsightly, it creates a trip hazard for those walking into the kitchen. We discussed this as a family and agreed to install hooks in the void beneath our staircase for bags to hang. This solution maximised this area and opened up our kitchen to be used as per its design.

Establishing boundaries around your stuff and your time will also prevent clutter from building back up.

Peter Walsh, a specialist in organisation design always says, "finish the cycle" and this is a good routine to embrace to help prevent clutter in your home. Write this down and stick it on your fridge because it's literally the one thing that will keep your house under control and clutter free. For example, washing the dishes is a cycle. If you're washing the dishes and sitting them in the drainer to dry, we all know that if we don't finish that cycle, it creates chaos. So instead, just dry the dishes, put them away, and wipe down the bench. It will then be a much neater, tidier space without the clutter. Another routine to get in the habit of is if something takes two minutes, just do it now. This will also help prevent clutter and chaos from building in your home.

Establishing boundaries around your stuff and your time will also prevent clutter from building back up. The best time to help children set boundaries around their stuff is about a month before their birthday or your cultural gift-giving season such as Christmas. Just say to your child(ren), "Gifts will soon be coming into the home. Let's have a look now and move stuff out, so we have space for the new stuff to come in." It's important to teach our kids how to be organised.

4. Become a conscious consumer

The final step in creating a minimalist home is to embrace the minimalist mindset and become a conscious consumer. In order to reduce the quantity of stuff you buy, consume and bring into your home, get in the habit of asking yourself the following questions each and every time you're about to buy something new:

Do I really NEED this item?

Question why you're buying the item in the first place. Are you only buying it because it's on sale, comes with a gift with purchase or because you're feeling flat and just want to buy something to make yourself feel better? If you can easily do without the item, leave it in the shop! It's as simple as that.

How long will this item LAST?

Another way to phrase this question is to ask, "what is the lifespan of this product?" The key here is to opt for quality, not quantity (such as classic clothing items rather than cheaper on trend pieces that last a few washes or at worst, shrink or lose their shape in the wash).

This question applies for smaller, seemingly insignificant items too: straws, tissues, paper towel, takeaway plastic cutlery, disposable coffee cups, an innocent bottle of water, nappies and wipes—what these items all have in common is that they're single use. You use them once and dispose of them ... forever! If you have the option to buy an item that can be reused several times rather than a single-use item, consider it a wise investment.

How OFTEN will I use this item?

Before you purchase a new item, think about how often, or even how long, you'll use it. If the answer is not much or not long, consider renting the item or doing without it altogether. You may even like to do some quick sums in your head and work out the price per use.

If you buy a $300 dress and wear it once, it's a $300 per wear item. If you buy a $300 dress and wear it 20 times while it's in your possession, it's a $15 per wear item. Traditionally men excel in this area better than women. Most grooms think nothing of hiring a suit for their wedding day yet how many brides do you know of who hired their wedding gown?

Of course, this doesn't just apply to clothing. A great example of an item used for a short amount of time is a baby bath. We bathed both our babies in the kitchen sink (with the dishes removed, of course!) for the first month or two and then transitioned to the regular bath, using a baby bath seat that we passed onto a friend when the boys outgrew it. Investing in a baby bath just seemed a bit ridiculous to us, especially when it would take up much of the space in our bathroom!

The question of how often you'll use the item is also applicable to bigger ticket items like car trailers, golf buggies, scuba equipment (I have a prescription mask but hire my buoyancy compensating device and tanks on the odd occasion when I scuba dive) and even boats or caravans that you may only use once or twice per year!! We hired a motorhome and had an amazing holiday when the boys were six months and three years old and more recently, we've enjoyed holidaying on the Murray River in a hired houseboat. These holidays were a fraction of the price they would have been if we owned the motorhome or houseboat and used them for a couple of weeks per year. What I'm trying to say here is don't get too precious about your belongings. When buying a new item that you know you'll only use occasionally, consider if

*Try not to use the fact
something has been
produced ethically and
sustainably as an excuse
to maintain high levels of
consumption.*

you can hire or borrow rather than buy, or do without the item altogether.

WHERE and HOW has the item been made?

The final question to ask yourself before buying something new is where and how the item has been made. Consider the distance from manufacturing location to market, conditions used in its manufacture and the materials used to make the item. Obviously, items made under ethical working conditions, using ethically sourced or produced raw materials (or even recycled materials) come out on top here, but it also pays to consider the distance to market. Try not to use the fact something has been produced ethically and sustainably as an excuse to maintain high levels of consumption. Eco-consumerism is still consumerism, after all.

Final thoughts

Embracing minimalism significantly lowers our environmental footprint. By decluttering our homes, setting up systems, routines and boundaries to manage our clutter, and becoming a conscious consumer, we'll control our stuff rather than having it control us and improve our happiness, health and financial situation.

Take action now!

It's over to you to take the steps required to improve your relationship with stuff and create your minimalist home. You can start today by completing the following three actions:

1. Declutter one area of your home, be it your wardrobe, bedroom or kitchen drawer. Move your items on straight away to a charity store. Don't delay because if the items are sitting in a box somewhere in your house, you or another family member could reclaim them back into your home.

2. Set up a system or routine for the area of your home you've just decluttered to prevent clutter building up again. Start by just sticking a "Finish the cycle" Post-it® note there to remind you to put everything back in its place.

3. If there's a larger, valuable item you own and rarely use, advertise it for sale today or donate it to a charity. It's likely best you hire the item occasionally rather than pay ongoing costs to insure, maintain and store the item.

Take before and after pictures of your decluttering efforts and share them on social media with the hashtag #sustainabilityinthesuburbs and tag me at @lauratrottadotcom so I can cheer you on!

CHAPTER 7
Lower the toxin load in your home

"You are what you eat, breathe and
apply to your body."
—LAURA TROTTA

From 1910 to 1987 canaries were used in underground coal mines to detect the presence of carbon monoxide gas. Their rapid breathing rate, small size and high metabolism compared to humans led the birds to succumb to toxic levels of gas before the miners, thereby alerting the miners to evacuate. Thanks to advancement in gas monitoring technology, canaries are no longer used in coal mines. However, "canaries" in our environment are sending us signals that levels of toxins are reaching concerning levels. These signals include things like fish kills when oxygen levels in rivers or lakes fall to insufficient levels (due to low flow) and failure of breeding populations in seabirds (due to starvation caused by bellies full of plastic). The same analogy applies in our own homes, where some of us, like the canaries in the coal mines, are finding that we too are reacting to toxins in our environment.

Growing up it was always the job of my two sisters and I to wash, dry and put away the dishes after every meal. Each time my elder sister washed the dishes she'd break out in painful, itchy and angry red hives all over her hands and wrists that would take days to subside. There was obviously a chemical in the dishwashing liquid that reacted with her skin. I initially thought she was using the hives as an excuse to get out of washing the dishes. But after the hives reappeared each time she did the job, I took on the role of dishwasher (albeit with dish gloves) and she switched to drying the dishes.

It's a concerning fact that many of the 40,000 individual chemicals permitted for use in Australia have never been assessed by the Australian Industrial Chemicals Introduction Scheme (AICIS), the chemical safety regulator. This is primarily because these chemicals were in use prior to 1990. Rather than detailed testing being the reason for approving their safety, they've been considered safe purely because they've been used for a reasonable period of time without known adverse effects. The long-term impact of such chemicals on our health, and the impact of different combinations of chemicals used in the home environment, are simply not fully understood.

The increased prevalence of chemicals in our homes and surrounding environments in recent decades has been linked to a significant increase in the prevalence of multiple allergic symptoms, asthma, rhinitis and eczema.[46] Like canaries in the coal mines who warned miners to evacuate to avoid succumbing to the toxic gases, the growing percentage of our human population with these conditions is a warning that our environment, including our homes, is becoming more toxic.

In my early twenties when I was suffering from debilitating migraines, I made a huge effort to clean up my diet and lifestyle. Room-by-room, product-by-product I created a toxin-free home, switching conventional products for naturally derived eco-friendly products, even making my own cleaning products using recipes from a bygone era. The process was slow and involved much research at my local library (there wasn't much information about toxin-free living on the Internet back then!) but the impact on my health was incredible. Within a few short months, I reduced my migraines from weekly to every few months and regained my health and zest for life in the process.

In this chapter, I'll share some tips for how you can reduce toxins in your home to improve your personal health, the health of our environment and help you save some money along the way.

Benefits of a toxin-free home

While I can list the lengthy benefits of a toxin-free home myself, I feel Jodie Wakefield one of the first graduates from my Home Detox Boot Camp, says it best …

I believe that reducing my kids' chemical load has calmed them down and enabled them to sleep soundly.

"My oven is clean for the first time in years, and I didn't have to evacuate the children and the dog during the process! But it gets better. My children S-L-E-E-P! My son just turned eight and never before had he slept through the night, every night. Needless to say, my five-year-old daughter was following suit with the bad habit. I believe that reducing their chemical load has calmed them down and enabled them to sleep soundly. My son would sneeze and sniffle every night getting into bed, which we treated with medicated nasal sprays (and believe me, I was tough on getting rid of dust mites with the information I had before the course—but it was all ineffective). He doesn't sneeze and sniffle in bed anymore, ever. The eczema on both of their skin has gone! It was proven to be the effect of the homemade bathroom products when my husband used naughty shop-bought soap in their bath one night and the eczema immediately came back. We no longer lather them in steroid cream and dose them up on antihistamines."

And Jodie's benefits also extended to the family's budget.

"In just one month I saved $450 in groceries (enough said—it's a miracle). I put it down to not buying any cleaning products and making soooo much of our food now. I had always done a LOT of home cooking (having a nut allergy in the house dictates this, plus my keen interest in healthy eating) but learning even more about the chemicals, preservatives, colours and additives from guest expert Bill Statham made me reassess our diet further than before."

Now while Jodie's results may not be typical for everyone (she didn't just complete my course, she also put the things she learnt into action), they do show what kind of transformation is possible when you detox your home.

A toxin-free home, where the cleaning, pest control and personal care products are all naturally derived, is a much healthier home. Reducing toxins in our home environment also improves our wider environment. Not only does reducing your consumption of chemical-laden cleaning, personal care and pest control products, reduce the overall demand (and manufacturing) of such products but you're reducing the amount of chemicals leaving your home via wastewater. Cleaner water leaving your home reduces greenhouse gas emissions and chemicals used in the wastewater treatment process.

A toxin-free home and lifestyle is also cheaper to maintain. The average Australian household uses over 30 different cleaning or pest control products. These can easily be reduced to a handful of six or seven natural products we can use for cleaning and pest control. This consolidation has a positive impact on your family budget and household clutter.

How to create a toxin-free home

I've guided hundreds of people to successfully detox their homes through my Home Detox Boot Camp and have distilled the process down to three simple steps:

1. decide where to start

2. learn to read product labels
3. make the switch to a less toxic alternative.

1. Decide where to start

Before you can actively remove toxins from your home, you need to understand what toxins are currently in your home and how you use them. Everyone doing my Home Detox Boot Camp starts this process by undertaking a home chemical audit. This is a simple process that involves taking a group photo of all the chemical products in different rooms of their house such as inside the laundry cupboard, bathroom cabinet and under the kitchen sink. They share the photo within our online community to document their baseline and gain accountability.

We've all fallen for the very successful marketing campaigns of cleaning and beauty brands over the years, and it's very easy to accumulate these products in our homes. It's commonplace to have anywhere between 30 and 60 products! There's no shame in having cupboards full of commercial cleaning, pest control and personal care products. What's important is the decision to break up with these products to improve your health and the health of our environment.

Once you have a clear idea of the toxins you currently have in your home, you can choose your starting point. You can start by product type, (for example cleaning, pest control or personal care products), or choose to focus initially on the chemical products used in one area of your home, such as the laundry, detoxing your home room-by-room. Cleaning

Toxins absorbed through our skin typically end up in our bloodstream and are transported to our organs.

products are a common place to start. But so too are personal care products, as the health benefit is greater given these products are applied directly to the skin.

Our skin is our largest organ and readily absorbs products that are applied. Unlike ingested toxins that are broken down by enzymes in our saliva and stomach, toxins absorbed through our skin typically end up in our bloodstream and are transported to our organs. Adult humans are reported to absorb 60% of what is applied to our skins and children are reported to absorb 40-50% more again. Products applied to armpits and genitalia are more readily absorbed than what is applied to skin on other parts of the body. For this reason, it pays to think twice about what menstrual products, deodorants, and personal lubricants you choose and be mindful of the nappy creams and ointments you apply to your children. If you or someone in your family is suffering from eczema or dermatitis it will make sense to start your home detox by reviewing all products that come into contact with the skin.

2. Learn how to read labels

Detoxing your home can sometimes feel like one big chemistry lesson and that's for a good reason. While you may have disliked chemistry at school, understanding what chemicals you're bringing into your home is the first step in becoming empowered to replace them with safer alternatives. Flipping over a product and checking out the ingredients list on the label, while daunting at first, soon becomes habit. It honestly won't be long until you start to recognise some of the lengthy ingredient names and return a product back to the shelf because it doesn't meet the standards you've set for YOUR home.

But what chemicals are particularly nasty and why? What are the "red flag" chemicals you should avoid wherever possible? Here are eight common household chemicals I recommend avoiding:

ALKYLPHENOLS

Alkylphenols are typically found in a wide range of everyday products from household cleaners and stain removers to pesticides, paints, hair care products, food packaging, plastics and even spermicides. They're particularly good at cleaning because they can tackle both water-based and oily messes at the same time. Unfortunately, because they're slow to biodegrade they tend to bioaccumulate in the food chain. That's no big surprise given there are many chemicals that do the same. What's most concerning, however, is that alkylphenols are linked to reproductive and developmental changes. In short, they can interact with cellular oestrogen receptors in the body and can create oestrogen displacement and havoc.

TRICLOSAN

Triclosan is an antibacterial and antifungal chemical that can be found in cleaning products and personal items such as antibacterial soaps, detergents, toothpastes, deodorants, facial cleansers, exfoliants, mouthwash and cleaning supplies. The chemical has been reported to affect the body's hormone systems, such as thyroid hormones, and consequently, may disrupt normal breast development. Due to public pressure and changing regulations, many major brands have quietly begun reformulating their products without triclosan; however, there are still many products out there that contain this chemical.

ALDEHYDES

Aldehydes, such as glutaraldehyde, are chemicals that may cause irritation when they are breathed in or come in contact with the skin. They can cause permanent damage to the eyes, ears, nose, throat and lungs. Formaldehyde is a lung and respiratory irritant and is also classified as a Group 3 carcinogen. Formaldehyde is commonly used as a preservative in cleaning products. It can also be found in cosmetics and personal care products such as cleansers, fingernail varnishes and hardeners, shampoos and conditioners, toothpastes and hair straightening solutions, household cleaning products such as carpet and rug cleaners, disinfectants, dishwashing liquids, floor cleaners and polishes.

BENZALKONIUM CHLORIDE

Benzalkonium chloride is a sensitiser especially dangerous for people with asthma or skin conditions such as eczema and chronic dermatitis. There is also a stated correlation between an increase in childhood asthma and the exposure to this chemical through household disinfectants, sanitisers and personal care products. Benzalkonium chloride is a common ingredient in commercial household disinfectant products.

Some ways you can avoid phthalates include ditching your perfume or air freshener and avoiding soft plastics where possible.

SODIUM HYPOCHLORITE

Sodium hypochlorite, also known as liquid bleach, is a strong oxidiser that can burn skin and cause eye damage, especially when used in concentrated forms. Mixing bleach with other household products, such as an acid, can be extremely dangerous and result in the production of toxic chlorine gas. Sodium hypochlorite is found in many commercial household disinfectants and bleach products.

PHTHALATES

Phthalates are endocrine-disrupting chemicals commonly used to render plastics soft and flexible. They are found in plastics, cosmetics, fragrances, household cleaners, baby care products, building materials, modelling clay, cars and insecticides. Phthalates enter the body by skin, ingestion, inhalation and medical injection. Some ways you can avoid phthalates include ditching your perfume or air freshener (the ingredient parfum is code for phthalate) and avoiding soft plastics where possible (this means not heating your food in plastic containers as well).

AMMONIUM HYDROXIDE

Ammonium hydroxide is used as a home sanitiser and is typically found in window cleaners. Commonly available ammonia that has had soap added to it is known as "cloudy ammonia". According to the World Health Organization and the United States Environmental Protection Agency, ammonium hydroxide is a carcinogen. It has also been linked with creating health problems with skin, liver, kidneys, lungs and eyes. The Environmental Working Group has stated a correlation with asthma, respiratory and skin issues as well.

DYES IN CLEANING PRODUCTS

Dyes in cleaning products are often unlabelled on the products' ingredient lists, but are comprised of several different chemicals, some of which are known carcinogens. Dyes in food and cleaning products have been linked to cancer and neurological problems, such as behaviour, attention and learning problems.

These chemicals really are the tip of the iceberg. To continue educating yourself about toxins and their health impacts I recommend downloading The Chemical Maze app on your smartphone. Developed by Australian scientist and researcher Bill Statham, the app enables you to search any chemical found in cleaning products, personal care products, or even food to see the potential health impacts. This app was developed to be a shopping companion, enabling users to review ingredient listings in products to inform a toxin-free purchase.

A word on greenwashing

Many products say that they're clean, green and organic, but the term organic is actually a chemistry term that simply means it contains carbon; it doesn't mean the product is toxin-free. To be certain you are indeed purchasing a naturally derived product that contains no nasty chemicals, you need to read and understand the label. If in doubt, opt for a certified organic product that contains ingredients that are produced without the use of synthetic chemicals, fertilisers or genetically modified organisms.

3. Make the switch or make your own!

The third step in detoxing your home is to replace commercial products containing toxins with a toxin-free alternative. A pro tip is to go plastic-free at the same time. There are some great eco-friendly brands that, in addition to being toxin-free, also have either no packaging or are sold in concentrate form. Plastic-free natural shampoo bars or powders are a great example. Switch to these products to reduce your chemical load and your plastic waste at the same time.

Switching to toxin-free products can also involve making some of your own products, which is not as difficult or time-consuming as you may think. If you're interested in making some of your own cleaning products, sign up for my free Green Cleaning Challenge at *cleancleaning.lauratrotta.com*. Over five days I share how you can use fresh lemons, white vinegar, baking soda, salt and olive oil to clean your home. My Home Detox Boot Camp obviously unpacks this step in great depth, sharing how you can make your own toxin-free products for many purposes in your home, such as laundry powders and liquids, mould removers and natural deodorants.

I know you may be thinking, "I don't have time to make my own cleaning products." If that's you, just take it slow by switching one product out a month. Or start by buying some more environmentally friendly and certified organic brands.

There is a lot of greenwashing out there so look for reputable eco brands that have won toxin-free or zero-waste awards.

There is a lot of greenwashing out there so look for reputable eco brands that have won toxin-free or zero-waste awards. Some good natural cleaning brands in Australia are Abode, Ecostore, Kin Kin, Koh and Resparkle.

If you're worrying that these products might cost more, let me put your mind at ease as there's a very cost-effective way you can use these products. Many health food or eco stores sell these brands in bulk. You can take empty bottles to the store and fill up directly there, saving a considerable amount of money in the process.

It's also not uncommon to believe naturally derived products don't work as well as their commercial counterparts. Perhaps you've used some eco cleaning or body care products in the past and you haven't been as happy with their performance as conventional ones. My advice is to keep trying and find those that don't just work as well, but outperform their toxic alternatives. There are more and more reputable eco brands on the market every year and unlike 20 years ago when some toxin-free products didn't perform, the bar has risen. In fact, some natural products clearly outperform their commercial alternatives. Mould cleaners are a classic example. Chlorine bleach is used in commercial mould cleaning products, but it doesn't actually kill mould spores; it just bleaches them white. Natural products like white vinegar, tea tree essential oil, and even clove essential oil are much more effective at killing mould spores.

Final thoughts

Detoxing your home is not just good for the environment and your family budget, it's a straight up health improver too. Twenty years has now passed since chronic weekly migraines set me on my home detox journey and I'm thrilled to report my migraines are only an occasional occurrence. If you or anyone in your household is suffering from any health condition, just give toxin-free living a go. You have absolutely nothing to lose, and everything to gain!

Take action now!

It's over to you to take the steps required to create your toxin-free home. You can start by taking the following three actions:

1. Do a home chemical audit by taking a baseline photograph of the products underneath your kitchen sink, in your laundry cupboard and bathroom cabinet.
2. Download The Chemical Maze App on your smartphone and look up the chemicals in three products you commonly use in your home.
3. Choose one product that you'll replace with a toxin-free alternative and make the switch today!

For other tips and recipes to detox your home take my FREE 5-day Green Cleaning Challenge at *cleancleaning.lauratrotta.com* or join my Home Detox Boot Camp at **homedetoxbootcamp.com**.

Share what changes you've made to detox your home on social media with the hashtag #sustainabilityinthesuburbs and tag me @lauratrottadotcom so I can personally congratulate and thank you.

CHAPTER 8
Work towards a net-zero home

*"I'd put my money on the sun and solar energy.
What a source of power! I hope we don't
have to wait until oil and coal run out
before we tackle that."*
—Thomas A. Edison

The summers of 2018/2019 and 2019/2020 were excessively hot in Australia. The country sweltered through heatwave after heatwave. Destructive and deadly bushfires burned in every state. My home city of Adelaide soared to 46.6°C (115.88°F) on 24 January 2019, the hottest temperature recorded in any Australian state capital city since records began 80 years ago. That same day, a town just a couple of hours drive away, Port Augusta, reached 49.5°C (121.1°F). This set an all-time temperature record for any Australian town. I remember this day well because I was working in the Adelaide CBD, conscious that as the temperature climbed, so too did the frequency of ambulance sirens in the city streets. Distracted by the ambulance sirens, I looked out my office window to see with horror, person after person fainting from heat exhaustion at the tram stop in Victoria Square. Heatwaves kill more people in Australia than any other natural disaster. South Australia's health authorities reported that 44 people received emergency treatment for heat-related illnesses on 24 January 2019.[47]

My children were home with a carer as it was school holidays. I'd almost returned home from work when the carer rang me distressed to say, "The power's gone out and it's really, really hot here. What do I do?" I arrived home to find that indeed, we along with the other houses in the street had no power, the outside temperature was 48°C (118.4°F) and the temperature inside our home was rising fast. Without air conditioning or a fan there was no cool airflow in the house, and we were all uncomfortable and getting very irritable.

I rang our energy provider and was advised there was an electrical fault in our suburb (not an uncommon experience in heatwaves) and that the power would be out for the next five hours or so. With that news we quickly left the house and headed to a nearby air-conditioned restaurant for dinner before spending the evening cooling down in the water at the beach. At 10:30 pm that night the temperature was still over 40°C (104°F) and hundreds of people were just sitting in the water at the beach because they either didn't have air conditioning at home, couldn't afford to run it, or like us, had power cut to their home. We returned home around 11 pm after the power supply had been restored. My exhausted kids fell into bed three hours past their bedtime and I quickly followed suit. The evening was unpleasant, yet I recognise how privileged we were to be able to afford to escape to an air-conditioned restaurant and have the option to cool off at the beach. Millions of people globally just don't have that luxury in heatwaves.

This is climate change. We're seeing:

» More extreme heatwaves, bushfires and droughts.

» Less rainfall overall, but when the rain does fall, it's in the form of extreme events that often lead to the kind of destructive flooding normally seen every 50 years or so (but is now happening twice or three times in one year).

» Massive storms blowing over power transmission lines and creating extended power outages.

» Sea levels that are rising.

» Mass migrations of people from drought-stricken countries moving to countries where there is more food or where they can find work.

Climate change is here. It's impacting every city and town, urban and regional area in the world. It's disproportionally impacting poor people in low-income communities and developing countries around the world due to their increased exposure and vulnerability.

So, what can you do about climate change?

South Australia is on track to have 75% of the grid decarbonised by 2025, 100% decarbonised by 2030, and 500% decarbonised by 2050.

One of the biggest actions you can take to address climate change is to reduce your own greenhouse gas emissions. The best place to do this is in your own home. Some of this work will be done for you if you live in a region with a proactive government that's actively working to decarbonise the electricity grid. South Australia, where I live, is on track to have 75% of the grid decarbonised by 2025, 100% decarbonised by 2030, and 500% decarbonised by 2050. This means that by 2050, the state will be actively exporting renewable energy to other jurisdictions within Australia and internationally. This also means that all 1.8 million South Australian residents will be using renewable energy to power their homes even if they don't have solar panels themselves. There are several states in Australia or internationally where the sole energy source for the electricity grid is still coal or gas. If you live in such an area, reducing your emissions at home is even more critical.

In this chapter I'll share some strategies on how you can reduce your greenhouse gas emissions and create a net-zero home; a home where the energy consumed is in balance with the energy generated.

The benefits of you creating a net-zero home are global but will also have direct benefit to you individually:

» You will significantly reduce your power bills and fuel costs. In fact, you can almost wipe them out entirely.

» You will become more self-sufficient and will have increased freedom from the rising costs of electricity and petrol.

» You can even set up your home such that when the power goes out in your suburb, like it did for me on 24 January 2019, your home can still have power. (While a home battery system may not be sufficient to run your air conditioning system, it will keep your refrigerator on, power some lights and run a ceiling fan.)

» Lastly, you'll reduce your green guilt and will be so proud that you're reducing your own emissions and doing your bit to help combat the climate crisis. You'll be part of the solution, not just one other household emitting greenhouse gas emissions and contributing to irreversible climate impacts that future generations will need to battle.

If you're struggling to sleep at night because you're conscious of your environmental footprint, this chapter is for you.

Let's now talk about the steps that you need to take to decarbonise your home. For some of these actions, you'll be tempted to wait—for more people to take action, for prices to fall, or simply 'until you're ready'. My request to you here is—if you can afford to, please don't wait. We don't really have time to wait for every government on Earth to get organised when it comes to climate action. The more of us who start where we have the most control (decarbonising our homes), the better placed we'll all be to create a better climate future.

How to create a net-zero home

1. Source renewable power

You don't need to own your own home to source renewable power; both renters and homeowners can switch to a green energy provider. Some states are already a fair way along decarbonising their electricity grids. If you live in one of these states, most of your power will already be coming from renewable sources. If this isn't you, you have a choice and that's to shop around and choose a green energy retailer.

If you do own your own home (or are paying a mortgage on one), choosing a green energy retailer is a good start, but the ultimate solution is to install solar panels and a home battery system on your property. The solar system will extract the energy from the sun during the day to power your household and put any excess power that you don't use into the grid. (You will get a small rebate for that on your electricity bill.) A home battery system combined with your solar system will enable you to continue to power your home through the evening and overnight using the solar energy you've stored in your battery during the day.

Both systems are an investment, and the cost may be a barrier for you. However, many governments offer rebates and finance options to spread the cost over many years, enabling you to still access the savings on your electricity bill without the large upfront cash outlay. These finance options may result in you being able to pay off your solar using the money

---

Choosing a green energy retailer is a good start, but the ultimate solution is to install solar panels and a home battery system.

---

you otherwise would have spent purchasing electricity, and it balances out that way. You also don't need to install both systems at the same time. You can start with solar and add battery storage later.

The statewide power outage in South Australia in 2016 left my home without electricity for four days. After my suburb's six-hour power outage during the January 2019 heatwave when temperatures topped 46.6°C, I decided that I not only wanted to embrace renewable energy to reduce my household's emissions, I wanted to become more self-sufficient and future-proof my home against blackouts.

When we built our home in 2017/2018, we did have some money put aside for solar. However, as our house build ran six months over schedule and we faced higher, unbudgeted costs from temporary accommodation, placing our belongings into storage and our pet cat into boarding, we moved into our home without solar. It was another year or two until we were in a position where we could take the first step to get the solar panels installed on our roof and another two years before we could install our home battery system. A $2,000 home energy grant from the South Australian government helped make our home battery dream a reality. Finally, six years after that four-day power outage I'd achieved my goal of decarbonising my home's electricity use and becoming more self-sufficient. It felt amazing.

I appreciate that solar panels and a home battery storage system is way out of most people's budgets. If it is in your budget though and you're yet to take the plunge, I would just like to say that with privilege comes responsibility. For prices to fall for everyone, it's going to take those people who have the privilege and ability to purchase these systems to do so for prices to reduce. If that's you, please take one for the team. The benefits to you are so worth it:

» You'll protect yourself against the rising cost of electricity.

» You'll have reliability of power so you can still turn on the light and cook your dinner during a power outage.

» You'll be proud that you're not only reducing your household's greenhouse gas emissions, but you're also putting renewable energy into the grid for others to access, as well as increasing the demand for solar and home battery storage systems so that costs continue to fall for everyone.

2. Electrify your household and transportation systems

The second step in creating a net-zero home is to electrify everything. Once you've tapped into a renewable source for power, you'll need to change all your appliances and systems to using this renewable electricity rather than energy from fossil-fuel sources such as natural gas, petrol and diesel. This involves electrifying all your appliances including your household's hot water system, oven, stove/cooktop, heating and cooling system, vehicle/s and even your lawnmower. Your goal is to have everything that you power in your house running off renewable electricity, so you have no greenhouse gas emissions.

The key here is to also prioritise using these systems, including charging an electric car, during the day when you can directly use electricity from your home solar system and not draw from the grid after sundown. Since installing our home battery and having our energy use data at our fingertips via an app, we've been able to see:

>> our total energy consumption
>> where our energy is coming from (solar, battery or grid)
>> the amount of power we have stored in our battery.

We run our dishwasher, washing machine, and even oven (where possible) during the day, when they're powered directly from the sun. When we do run these appliances in the evening after sundown, we're more likely to drain our battery and then need to draw electricity from the grid overnight.

I've mentioned moving to an electric vehicle here and, again, I recognise this is another considerable investment. Like home solar and battery systems, many governments offer rebates to encourage electric vehicle uptake. At the time of writing this book I've had an electric vehicle on order for over six months and, thanks to ongoing supply chain disruptions, it will likely be another six months until it arrives. In the interim, we have a hybrid car that offers much

Reducing your home energy consumption is a critical step in your net zero journey.

better fuel efficiency and lower greenhouse gas emissions than a standard internal combustion engine vehicle. Hybrid and electric vehicles will continue to become more accessible and affordable in coming years and I encourage you to do your sums, investigate what rebates are available to you, and make the move as soon as you can.

Range anxiety is often given as a reason for not making the switch to an electric vehicle. Most models can now travel at least 400 km (249 miles) before needing to be recharged. If you're a city or a town dweller, that range is more than adequate to run your errands, do the school run and travel to and from work if you need your vehicle for commuting. I can't wait to drive the 800 km from Adelaide to Melbourne in our electric vehicle and charge up once or twice along the way. There are now charging stations every 200 km or so in towns that we'd be stopping in anyway to break up the drive. With the charging process taking around 30 minutes at a charging station that just gives us all a bit of extra time to stretch our legs and refresh before continuing the journey.

3. Reduce your energy consumption

Regardless of whether you're a renter or a homeowner with or without rooftop solar, reducing your home energy consumption is a critical step in your net-zero journey. This is important because it helps you to reduce the electricity you draw from the grid (if you don't have solar) as well as maximise the renewable energy you're able to put back into the grid for others to use (if you have solar).

You'll be surprised how much energy use you can reduce in your home through small actions. When volunteering as a home sustainability auditor in my local community, a town mostly comprised of renters, I was able to help households reduce their energy consumption and power bills by up to 40% in some cases, just by taking a series of simple steps such as the ones I'm about to share with you. Don't underestimate the impact of what small changes like these can make.

In an average Australian home, heating or cooling our living space makes up just under 40% of household energy use, followed by water heating at 25%. Actions to reduce the use of your air conditioner in summer and heater in winter, as well as your consumption of hot water, can make a significant positive dent in your household energy consumption and your power bill. Here are three areas you can focus on to reduce your household energy consumption:

1. Optimise your household heating and cooling

The most impactful home energy tip is to set your heating thermostat at 18-20°C (64.4-68°F) in winter and 24-27°C (75.2-80.6°F) in summer. Every degree warmer (heating) or cooler (cooling) than this adds an additional 10% to the energy use and running cost of the appliance. Think twice before lowering or raising your thermostat. Can you turn on a ceiling fan if you're hot or put on more clothing if you're cold instead of turning on your air conditioner or heater? By investing in insulation in your walls, ceiling and even floors, you can effectively maintain the temperature in your home and are less likely to need your heater or air conditioner.

On those days when you do need to turn your heating or cooling on, practise 'zoning' and only heat/cool rooms that are occupied in your home. It really makes no sense to heat or cool your laundry if you're only in there for five minutes a day! Get in the habit of shutting the doors to rooms you don't use during the day, such as bedrooms. You can easily open them again in the evening to warm/cool them down prior to going to bed.

2. Show your hot water service some love

I'm guessing you've never given much thought to your hot water service (unless it stopped working and you needed to replace it) but this workhorse needs a little TLC every now and then. Water heating typically makes up a staggering 25% of our home energy bills! This means if you want to reduce your home energy use, you'll need to get to know your hot water service a little better.

Many homes with storage hot water services have their thermostats set too high. For safety reasons, your thermostat should be set to a minimum of 60°C (140°F) to prevent growth of harmful Legionella bacteria within the unit. Any temperature higher than this means that energy (that you're paying for) is used unnecessarily. To determine if your hot water service is set at the optimum temperature use a home energy thermometer to test the temperature of water leaving the unit. Some hot water services let you do this directly at the unit, but for others you'll need to do this inside your home at the closest hot water tap to the unit. This might be in your bathroom, laundry or kitchen. Run the hot water over your

Much heat in winter can be lost from the metal pipes attached to your hot water service.

thermometer for a few seconds and read the temperature. If the water temperature at the unit is higher than 65°C (149°F) reduce the temperature setting to 60°C. If you're measuring the temperature at a tap inside your house, you'll need to allow for temperature loss in your pipes. You want to aim for a temperature here of 50°C (122°F). If it's significantly warmer than 50°C, lower the thermostat of your unit.

Just like heat can be lost when piping hot water around your home, much heat in winter can be lost from the metal pipes attached to your hot water service. This is an issue as the more heat lost, the harder your hot water service is working to heat up more water, and the more energy it's using. The amount of heat lost can be significantly reduced by insulating pipes connected to your hot water service with lagging— foam tubing available from your local hardware store. To use, cut along the length so you can easily feed the lagging around the pipe. Secure with tape if required.

It pays to regularly check that there are no drips from your unit, including from the relief valve from your heater. As a rule of thumb, if you place a bucket under the relief valve and it fills in a day, your valve needs replacing.

It's one thing to ensure your hot water service is running efficiently, it's another to reduce the amount of hot water (and by default, energy you use to heat hot water) in your home. One action you can take here that can make a big impact is to wash your laundry in cold water. It's the mechanical action of your machine and detergent that cleans your clothes— washing in warm water uses unnecessary electricity.

Reducing the amount of hot water you use in the shower also makes a big difference to your home energy consumption. I cover actions on how to reduce hot water use in the shower in Chapter 9, Be more water smart at home.

3. Optimise your fridge and freezer

Our fridges and freezers are continually running and, therefore, continually using electricity. Fridges and freezers can be costly to run and unsurprisingly, the cooler their temperature, the more electricity they use! It may seem obvious, but once food is frozen, it's frozen. It's no more frozen at -15°C (5°F) than it is at -22°C (-7.6°F) BUT the electricity used by your freezer to run at -22°C is considerably more than at -15°C. To measure the temperature inside your fridge and freezer, place a home energy efficiency thermometer inside, close the door and open again after five minutes. Ensure your fridge's thermostat is set to maintain a fridge temperature between 3°C and 5°C (37.4°F and 41°F) and your freezer to between -12°C and -8°C (10.4°F and -0.4°F). Your food will still remain cool or frozen, but you'll save hundreds of dollars (and greenhouse gases) over the years by maintaining your fridge and freezer within these temperature ranges.

Do you run more than one fridge or freezer in your home? A survey of my greenHOUSE Home Energy Blitz participants showed that two thirds operate two or more fridges and freezers. That's pretty common in Australia and I'm also included in that number; we have one fridge and one chest freezer that are used continuously. I honestly couldn't survive without my chest freezer as I do a lot of batch cooking and freezing of family meals, stocks and soups. This makes it possible to dish up nutritious homemade meals during the week when I'm working and running the kids around to extracurricular activities. It also provides contingency during the tough times, such as the weeks following my wrist surgery for carpal tunnel when I didn't have the strength to chop any vegetables or lift pots and pans. This is an example where I've weighed up the cost to own and operate something against the convenience it offers.

You might be in a similar situation and have made a conscious decision to run a second fridge or freezer or you may not have given it much thought. Either way it's worth just touching on the impact of this choice. What's your second or third fridge or freezer used for?

It is a drinks fridge? A mainly empty fridge that picks up overflow from your main fridge? A fridge you only turn on when entertaining?

You'd be surprised just how many homes I've audited that have a drinks fridge, often located outside (in direct sunlight sometimes!), just for the luxury of enjoying a cold beverage in the evening after returning from work. Now, like my chest freezer, this is a lifestyle choice the homeowner has made, but a choice that's good to question. **A second fridge typically adds around $200 to a household's annual electricity bill.** That's a lot of beer!

Given that many people use old, inefficient fridges as second fridges too, this value can easily be a lot higher. If you run a separate drinks fridge, consider keeping a small supply of

Offsetting is a bandaid fix to emissions. It doesn't address the root cause of climate change.

drinks in your main fridge and just top up as required. Do you really need an entire fridge of drinks on standby? Maybe you might decide to turn your second fridge off and just run it when entertaining.

By actioning the above tips to optimise your home heating and cooling system, hot water service, fridge and freezer you'll be able to make significant reductions in your electricity use, greenhouse gas emissions and power bills.

Can't I just offset my emissions?

Now you might be thinking, do I really need to do this work to reduce my home energy consumption and make these investments into solar and perhaps a home battery system or electric vehicle? Can't I just offset my emissions? Well, yes you could, but offsets take years to remove carbon from the atmosphere. We need to reduce our direct greenhouse gas emissions at the source, which is our homes and our transportation systems, right now. Offsetting is a bandaid fix to emissions. It doesn't address the root cause of climate change, (the extraction and burning of fossil fuels to power our lifestyle). The sooner we reduce our emissions of greenhouse gases, the sooner we protect not only the lifestyles of future generations, but their mere survival.

Final thoughts

Decarbonising your home and lifestyle through sourcing renewable energy, electrifying your household and transportation systems, and reducing your energy consumption will effectively create your net-zero home and reduce your footprint on our planet. It will also help to stave off some of the worst projected impacts of climate change, and that's worth investing both time and money to achieve.

Take action now!

It's over to you to take the steps required to create your net-zero home. You can start by taking the following three actions:

1. Set your home heating/air conditioning thermostat to 18°C-20°C (64.4°F-68°F) in winter and 24-27°C (75.2°F-80.6°F) in summer.

2. Measure the temperature of your hot water service and ensure it's set to around 60°C (140°F) to maintain the health of your water supply while minimising unnecessary energy consumption.

3. Measure the temperature inside your refrigerator and freezer and adjust your fridge thermostat to between 3°C and 5°C (37.4°F and 41°F) and your freezer thermostat to between -12 and -18°C (10.4°F and -0.4°F) if they're cooler than these temperatures. Retire one of your spare fridges or freezers from everyday use, using it just when entertaining or in busy periods.

Share what changes you've made to decarbonise your home on social media with the hashtag #sustainabilityinthesuburbs and tag me @lauratrottadotcom so I can personally congratulate and thank you.

You can purchase a home energy efficiency thermometer at **lauratrotta.com/product/home-energy-thermometer.**

For other tips to reduce your household energy consumption, take my FREE 5 Day Home Energy Challenge at **5day.lauratrotta.com.**

CHAPTER 9
Be more water smart at home

---

*"When the well is dry, we will know
the worth of water."*
——BENJAMIN FRANKLIN

When my sons were very young, we enjoyed a motorhome holiday, driving the 3,000 km round trip from our home in outback South Australia to Melbourne, Victoria to visit my sisters and meet their newborn baby girls. It was quite an adventure for so many reasons, but one memory in particular stands out for me. We were parked in a caravan park in suburban Melbourne one evening when the heavens opened with a heavy downpour of rain. Unlike everyone who raced undercover, my eldest son Matthew, who'd just turned three at the time, ran straight out into the rain and played in the puddles without a care in the world. Being a desert kid, it was his first experience of rain. He was so thrilled that he could stomp in muddy puddles, just like Peppa Pig. The joy of rain outweighed the inconvenience of him being cold and wet. It was a simple reminder to me of how the simple things in life can bring the most joy.

Water is joyful, water is essential for life, water is life. Water is also our most precious resource. Without adequate fresh water, human life and the ecosystems that support us simply can't survive. But sadly, as climate change continues to impact rainfall patterns and other human factors impact water quality, the world's freshwater resources find themselves under immense pressure and are expected to decline. Scientists estimate that:

» By 2025, half of the world's population will be living in areas where water is scarce.

» By 2030, around 700 million people could be displaced by intense water security.

» By 2040, roughly one in four children worldwide will be living in areas of extremely high water stress.[48]

Even in countries with adequate resources, water scarcity is not uncommon. From Broken Hill in Australia to Cape Town in South Africa, many towns and cities around the world have been on the brink of running out of water. The quality of our freshwater resources is also impacted by pollution and the presence of forever chemicals, such as perfluoroalkyl and polyfluoroalkyl substances (PFAS), in our environment. Every human has a right to access safe water and sanitation and we cannot, and should not, accept water scarcity.

Water is a finite resource; our planet won't miraculously make more water as the human population soars. We, therefore, all need to become smarter with how we use this valuable resource and reduce water consumption in our homes. Conserving water, even in areas where water is plentiful:

» helps reduce the need to build dams or increase their capacity

» protects river health by reducing the volumes of water extracted

» decreases the volume of water and wastewater requiring treatment (which in turn, reduces greenhouse gas emissions associated with water pumping and treatment).

The amount of water we use has a far-reaching impact on the environment, past the borders of our town, city, state and even country. To ensure adequate future supplies of fresh, clean water, we need to value and treat it as the precious resource it is. Residents of towns and cities where quality water is literally on tap need to stop taking water for granted and using it like it's endlessly abundant, because it's not. Regardless of wherever you live you can smarten up how you use water to ensure there's enough to go around now and in the future. By reducing your water consumption and creating a water-smart home, you'll significantly reduce your water bills and effectively drought-proof your home, ensuring your lifestyle isn't impacted by future water restrictions.

I recommend increasing your household water storage by installing rainwater tanks.

How to create a water-smart home

So, how can you create a water-smart home? Essentially, you'll need to focus on the following three areas:

1. increase water capture and storage
2. reduce water consumption
3. increase water reuse and recycling.

Let's unpack each of these before diving in further on how to reduce water consumption in the home.

1. Increase water capture and storage

Building our home in the driest major city (Adelaide) in the driest inhabited continent (Australia) was a lesson in the high demand for water capture and storage systems. Within hours of our water tank being delivered and placed at the front of our house, it was stolen. I'm not sure if people are driving around the suburbs just looking for the opportunity to sneak a 2000 litre water tank into the tray of their ute, but the value of a tank and the water it stores is obviously a reason for some people to resort to theft!

While I don't condone stealing, I do recommend increasing your household water storage by installing rainwater tanks to capture all the rain that falls on your roof to use in your home. Rainwater tanks, combined with a water filter, have the added benefit of providing you with a very high-quality drinking water for next to no cost beyond the original outlay.

2. Reduce water consumption

You can't create a water-smart home without becoming savvier with how you use this precious resource. Reducing your water consumption requires both behavioural change and some tweaks to the appliances and devices that use water in your home to ensure they're operating efficiently.

3. Increase water reuse and recycling

Increasing the volume of water you *reuse* in the home is a key step in reducing your overall consumption. This involves adopting the mindset of not discarding any water unless it absolutely cannot be used again. Simple changes like bathing your children in the same bath of water (not necessarily at the same time), pouring leftover water from your family's reusable water bottles into the kettle at the end of a school day to use in your next cup of tea or in cooking, or diverting wash water from your laundry load to other uses, such as flushing the toilet or watering the garden or lawn, will enable you to reuse significant volumes of water. Prior to diverting any wash water from your laundry, check with your local council to ensure that any greywater systems you install comply with regulations and employ an accredited plumber to perform the task.

Three ways to reduce your household water consumption

There are many areas where you can save water in your household. But for the purpose of making fast progress in this area, let's focus on the three areas where we consume the most water. Inside the home, the shower is typically the biggest water user, clocking up around one third of indoor water use in an Australian home. This is closely followed by the toilet and laundry, which each account for approximately 25% of total indoor water use.

1. Shower with less

When volunteering as a sustainability auditor in my local community, one audit stood out. On arrival at Lorraine's home, she declared at the outset that I could suggest improvements, but she was not going to reduce the duration of her shower. Her shower was her "me time" and she loved a long, hot shower every night. Lorraine wasn't unique in her request—there were many households where I was told, often by the woman of the house, that the shower was a "no go" zone.

I didn't want to be the bearer of bad news for Lorraine, but having a long hot shower is a double whammy in the eco stakes. It hits your household's water and energy consumption hard by using the most hot water out of all household activities. Keeping your showers under three minutes and installing a water-efficient showerhead are keys to achieving significant water reductions.

You can shower with less by attaching a three or four-minute shower timer to the wall in your shower and finishing your shower before the timer runs out.

Luckily for Lorraine, she was open to having a water timer in her shower and before too long she'd reduced the duration of her showers. She also agreed (reluctantly at first!) to change her showerhead over to a water-efficient model and to testing the showerhead for one week before she made a decision to either keep it or switch back to the high flow model. When I checked back in with Lorraine a couple of weeks later, she was pleased to report that the new showerhead didn't impact too much on her shower enjoyment. What she did realise though when her next electricity bill came was that her family had reduced their energy use (and bill) by around 25%!!

You can shower with less water by attaching a three or four-minute shower timer to the wall in your shower and finishing your shower before the timer runs out. You may need to do what I do to keep your showers to this duration and that's turn the water off while washing your hair, turning it back on again when it's time to rinse. Replacing your showerhead with a water-efficient model can help to reduce your hot water use in the shower by around half. An inefficient showerhead can use between 15 litres and 25 litres (4 gallons and 6.6 gallons) of water every minute, whereas an efficient WELS 4 star rated one uses as little as 5 litres (1.3 gallons) every minute. The reduction in hot water means less energy is needed for water heating, saving you hundreds of dollars each year on your energy and water bills.

If you don't know whether your existing showerhead is water efficient, the easiest way to determine this is to measure its flowrate using a bucket with litre measurements on the side and a stopwatch. Have someone hold the stopwatch (or your smartphone if you're using the stopwatch feature) so you don't juggle it and your bucket at the same time. Turn the water tap on full volume and as soon as the person holding the stopwatch tells you to start, place the bucket immediately under the showerhead. Remove the bucket after 10 seconds and turn the tap off. Measure the litres of water collected in your bucket and multiply this volume by six to discover the flowrate of your showerhead.

Ideally you want your flowrate to be less than 9 litres (2.4 gallons) per minute.

If you don't already have a water-efficient showerhead installed, you can purchase one from your local plumbing supplier or hardware store. It might take a little getting used to at first so give yourself one week to get used to the change in flow. At the same time, install a shower timer, and really try to keep the duration of your shower less than four minutes. Like Lorraine, after a few weeks, you won't notice the difference until your water and energy bills arrive and then you'll be absolutely thrilled!

If you have young children, bath them together and if you have a baby, consider bathing them in the sink, rather than a large baby bath. You'll save yourself the cost of a baby bath that you'll only use for a short while too.

2. Flush with less

Composting toilets have been installed in rest areas throughout many remote areas in Australia. These water-less toilets require you to sprinkle some dry material that contains microbes onto your waste before you close the toilet lid. The microbes and airflow then work to compost the contents of the toilet. Composting toilets have revolutionised toilet facilities in remote areas, replacing pit toilets which are renowned for having a very strong odour and attracting flies. Composting toilets aren't just for highway toilet facilities in remote areas, they can be installed in your home too. The result being that you don't need water at all to flush the toilet. This can reduce your total household water use by up to one third!

If the idea of a household composting toilet repulses you, you'll need to look at other ways to reduce the volume of water you're flushing down the toilet as well as the type of water that you're flushing. Toilet flushing is one of the highest areas of water use in an average home, so it presents a prime opportunity for water conservation:

» In a home with older toilets (installed between 1980 and 1992), an average flush uses about 13.6 litres (3.6 gallons), and the daily use is 71.2 litres (18.8 gallons) per person per day.

» In a home with high-efficiency toilets (HETs) with an average flush volume of 5 litres (1.3 gallons), the daily use is less than one third at 26 litres (6.9 gallons) per person per day.

Dual flush toilets are a type of HET with a full flush and a half flush capability. The average flush volume of a modern dual flush toilet is 4 litres (1.1 gallons) or less. The oldest toilets can use more than 30 litres (7.9 gallons) per flush![49]

*A leaking water cistern
can waste approximately
16,000 litres
(4,227 gallons) of
water per year.*

If upgrading your toilet is out of your price range for now, get in the habit of only flushing when necessary. The common saying "It it's yellow let it mellow, if it's brown flush it down!" holds very true here.

It's one thing to be mindful of how often you flush your toilet; it's another to know whether or not your toilet is leaking water. A leaking water cistern can waste approximately 16,000 litres (4,227 gallons) of water per year. This equates to around $55 on your annual water bill (using the average price paid by Australian households of $3.46 per kilolitre).[50] It pays to regularly check your toilet cistern for leaks. Most of these you'll never hear; it's only the significant leaks that create a hissing noise or present themselves by a constant stream of water flushing. A slow, barely visible leak can waste more than 4,000 litres (1,057 gallons) per year. Visible, constant leaks can waste more than 96,000 litres (25,360 gallons), costing you around $330 per year.

To check if your toilet cistern is leaking, turn the tap feeding the toilet off and place a few drops of food colouring in the cistern. If the colouring shows in your toilet bowl after 15 minutes you have a leak that requires attention. Call a registered plumber to address any leaks and while they're in your home, ask them to check all taps to make sure the washers are intact and there's no leaks. A tap leaking at the rate of one drip a second wastes more than 12,000 litres (3,170 gallons) of water per year.

The final action to reduce the volume of treated, potable water being flushed down your toilet is to have a registered plumber connect your rainwater tank to your toilet. This will save considerable money off your water bill and reduce the greenhouse gas emissions associated with treating water to potable standards and pumping it to your home to simply flush your waste. If you think that using rainwater to flush your toilet is a waste of a quality water resource, you may be interested in some research done by a team of environmental engineers from Drexel University in Pennsylvania, USA. This team showed that if homeowners in Philadelphia, New York, Seattle and Chicago, were able to collect and store the rainwater from their roofs, they could flush their toilets without having to use a drop of municipal water.

3. Wash with less

The laundry is a great place to not only reduce water consumption, but to harvest greywater for other uses in your home and garden. In our previous home in Roxby Downs, we installed a hose that diverted water from the rinse cycle of our washing machine onto our garden. We used a natural, eco-friendly laundry detergent, so there wasn't an issue of contaminating our soil or plants, and this enabled us to water our garden in the harsh desert climate where water prices were at a premium.

You can improve water efficiency in your laundry by washing only full loads. If you have a top loader, adjust the water level so it's appropriate for the size of the load; don't use a full load of water for a quarter load of laundry. When it comes time to upgrade your washing machine, choose a water-efficient model. Front loaders are the best option here; however, some efficient top loaders are now also available. If you choose a front loader, you'll not only save water but you'll save detergent as well.

Final thoughts

Increasing your household's water capture and storage, reducing your water consumption and increasing water reuse and recycling significantly lowers your environmental footprint and helps to preserve Earth's most precious resource, fresh water. Once you get started creating your water-smart home, you'll sleep well at night knowing that you're saving money, drought-proofing your home and living lighter on our beautiful planet.

Take action now!

It's over to you to take the steps required to make your home water smart. You can start by taking the following three actions:

1. Using a bucket and stopwatch, measure the flowrate of your showerhead and purchase a water-efficient showerhead if your flow rate is greater than 9 litres (2.4 gallons) per minute.

2. Install a 3 or 4-minute shower timer and make a real effort to lower your shower times (and get everyone else in your household to try too). You can purchase a shower timer at *lauratrotta.com/product/shower-timer*.

3. Check your toilets for leaks using the food colouring method, check your taps for drips and leaks and replace washers where necessary.

Share what changes you've made to make your home water smart on social media with hashtag #sustainabilityinthesuburbs and tag me @lauratrottadotcom so I can personally congratulate and thank you.

For other tips to reduce your household water consumption, collect your FREE cheat sheet with 30 *Simple Ways to Reduce Water Use In and Around Your Home* at **watersaving.lauratrotta.com.**

PART 3

EXPAND YOUR IMPACT

CHAPTER 10
Raise eco-conscious kids

*"Kids who are more connected to nature today will
be more inclined to conserve it tomorrow."*
—LAURA TROTTA

Growing up in regional Victoria in the 1980s and blessed with
the Gippsland Lakes, (Ramsar Wetlands of International
Importance) and the Victorian High Country literally on
our doorstep, much of my free time and family holidays
were spent outdoors. Play was outdoors too. If my sisters
and I weren't hanging upside down from the monkey
bars in our local playground or roller skating through our
neighbourhood, we could be found climbing the trees or
looking for gnomes in our garden. We had so much to do
outside, we simply didn't need many toys.

My early childhood in suburban Melbourne wasn't overly
different; riding our tricycles in the backyard, making mud

pies and playing hide and seek were all in a usual day's play. Screen time was non-existent and creative play and dress ups using my mother's wardrobe castoffs rather than a purchased child's costume were in fashion. Of course, my sisters and I longed to have an Atari video game device like some of our friends, but we weren't overly upset when the answer was "no". We were having a lot of fun playing together outside.

While at university in the mid-1990s, I observed that a large proportion of students in the environmental engineering degree I was studying also grew up in the country. They'd spent their childhood weekends playing down the local creek, picnicking in the countryside or helping out on their family farm. My city-raised classmates were typically the outdoor adventure types who grew up camping or bushwalking with their families or with local Scout or Guide troops. Despite being raised in the large city of Melbourne, they had connected with nature from a young age and didn't value it any less than the country students. It wasn't surprising then that many weekends and holidays in my late teens and early twenties were spent hiking mountains, cross-country skiing, or camping in the great outdoors with my university friends. It's also not surprising that these friends now all hold environmental and sustainability leadership positions in the corporate and not-for-profit sector in Australia and abroad. The deep connection they developed with nature as children was so strong. It still motivates every day of their working life to protect our environment for generations to come.

All aspects of our society—finance, politics, insurance, education, medicine, engineering and science, the arts and even defence—need leaders who value and care for our environment.

Developing a love and respect for nature as a child makes us more likely to value its conservation in future. Whether or not you are a parent, aunty, uncle or teacher, you have a role to play in nurturing the next generation of eco-conscious kids to become adults who care and advocate for our planet. This is a way that you can amplify your impact.

However, the world has changed significantly within a generation and we're raising today's kids in the digital, not analogue age in which many of us grew up. It can be more challenging to connect children to nature at a young age as we're competing with technology for their attention. Add increasing mortgages and cost of living pressures as well as grandparents unable to help with childcare, overstretched time-poor parents can be forgiven for dropping their guard where screens are concerned.

In order to raise a generation of eco-conscious kids with the skills and desire to be conscious leaders of the future, we need to:

» embrace the benefits of technology while also
» managing the challenges that screens present in connecting our children to nature, to each other and to us.

In this chapter, I'll share tips and strategies on how you can nurture the next generation.

To grow up healthy,
kids need to sit less and
play more.

Benefits of raising eco-conscious kids

Raising eco-conscious kids has benefits for children, adults and for our environment as a whole.

It improves the physical and mental health of our kids.

To grow up healthy, kids need to sit less and play more. Raising eco-conscious kids who enjoy spending active time in nature builds lifelong habits that reduce obesity and strengthen immune systems. Playing outdoors in nature also develops our children mentally, socially and physically. Through playing in nature and having the freedom to explore their world, we raise happier, calmer and healthier kids.

It also improves the ability of kids to solve the complex environmental issues of the future.

Reading sustainability-themed books to children, engaging them in creative play or fostering a love of maths and science improves their education experience and gives them a broader knowledge of the world around them.

Raising eco-conscious kids also costs less.

Eco-conscious kids are happy with fewer toys, technology and gadgets as they enjoy outdoor play, eco-play and community resources such as libraries and playgrounds. Raising eco-conscious kids also results in fewer medical bills thanks to a healthier composition, stronger immune system and lower probability of developing chronic health impacts as they age.

Raising eco-conscious kids also improves our relationships and connections with each other and within our families.

Eco-conscious families are less likely to have a disconnected tween or teen constantly distracted by gaming or social media. Through setting screen time limits for our kids and creating time to connect as a family, we can strengthen family relationships and build more resilient human beings.

Finally, raising eco-conscious kids develops leaders of the future who are more likely to advocate for our environment. Care for their environment and each other are important traits to have in childhood as well as adulthood.

The benefits of raising eco-conscious kids are worth the effort involved. Let's step up together to amplify our impact!

Five ways to nurture eco-conscious kids

There are five core ways in which we can nurture our next generation by raising eco-conscious kids. These strategies are relevant if you have children in your life, regardless of whether you're a parent, aunty, uncle, teacher, dance teacher or sporting coach.

1. Switch screen time for green time

In 2014 New York Times reporter Nick Bilton famously asked Steve Jobs, "So your kids must love the iPad." Jobs answered, "They haven't used it yet. We limit how much technology our kids use at home."

Despite building one of the world's biggest tech companies and earning billions of dollars from the success of devices like the iPod, iPad and iPhone, Jobs knew then what we know now. Screens are designed to be addictive and that too much technology is not the best for a child's optimum development.

Thanks to the COVID-19 pandemic and its resulting lockdowns, screen addiction in both children and adults is at record levels. An international study published on screen time use among three to seven-year-old children found that, on average, children engaged with screens more than 50 minutes more during the pandemic than before. This was largely driven by increases in screen use for entertainment purposes (nearly 40 minutes) as well as educational apps.[51]

Whether we love it or loathe it, our kids will inherit a digital world.

According to Dr Kristy Goodwin, one of Australia's leading digital parenting experts:

> "Digitally amputating our kids is not the solution. Banning it, avoiding it, making it toxic or taboo isn't helpful for our kids. What we need to do is to teach kids healthy ways to use technology that are congruent or aligned with how they develop, and at the same time balancing screen time and green time."

Technology has many positive benefits:

» it can allow kids to communicate in new ways

Used the right way for the right amount of time, technology can certainly help our kids.

» it can allow them to collaborate

» it can allow them to access information.

Used the right way for the right amount of time, technology can certainly help our kids. But, like anything, if it's used excessively or inappropriately, it can derail their development. According to Dr Goodwin, about 85% of a child's brain architecture is established in the first three years of life. If children are using screens for disproportionate amounts of time, they may not be able to meet some of their basic developmental needs, like physical activity, play, language, forming relationships and sleep.

Screens if used for too long, have the potential to displace important developmental priorities. There are also potential risks to children's physical health. Excessive use of screens has been linked to increased cases of myopia (nearsightedness) in children, noise-induced hearing loss because of incorrect use of headphones and impacts to musculoskeletal health. Conditions like 'tech neck' and repetitive stress injury from children playing for too long on screens are becoming more apparent as well.

It's not just our kids' physical health that's impacted by screens.

A 2021 study (Paulich et al.) on screen time and early adolescent mental health found that more screen time is moderately associated with worse mental health, increased behavioural problems, decreased academic performance and poorer sleep, but heightened quality of peer relationships.[52]

How much screen time is too much?

The Australian 24-hour movement guidelines were developed from systematic reviews of the evidence about the effects of physical activity, sleep and sedentary time (including screen time) on children's development, health and wellbeing. The guidelines recommend:

» No screen time for children younger than two years.

» No more than one hour per day for children aged between two and five years.

» No more than two hours of sedentary recreational screen time per day for children and young people aged five to 17 years, not including schoolwork.[53]

Like many households our tweens have frequent days when they break through the recommended two-hour barrier. It takes a real, concerted effort to manage their screen time, especially when we're working from home on screens at the same time and they're quick to notice the hypocrisy.

So how can we easily reduce our kids' screen time?

Dr Goodwin recommends that parents come up with a screen time limit that's right for your child as every kid has a different tipping point for how much it affects them, and how it affects them to come off a screen. She suggests focusing on what your child is doing with the technology, rather than how much time they're spending on their device:

» Is it leisure?

» Is it learning?

» Is it active or passive?

Dr Goodwin also suggests looking at when screens are used. The time of day when kids use screens can have a direct impact, particularly on their sleep and attention levels. Ultimately, limit your child's use of screens, particularly backlit devices such as tablets and smartphones, 90 minutes before sleep. This is because they emit blue light, which suppresses the body's production of melatonin. And kids are more susceptible to this as they need melatonin to be able to fall asleep quickly and easily. The use of backlit devices before bed and nap time can also delay the onset of sleep and over time these sleep delays can accumulate into a sleep deficit.

Also consider how kids are using screens; are they developing healthy habits that won't damage:

» their vision

» their hearing

» their posture or

» their health?

Dr. Goodwin says that looking at these four facets together gives us a more comprehensive picture than simply ticking a box saying, "Yes, they've only had an hour of screen time." Not every kid likes coming off screen time abruptly and many will have a techno-tantrum as a result. To avoid the techno-tantrum a positive transition activity is needed because their brain is getting a burst of dopamine when they're on a screen.

Not every kid likes coming off screen time abruptly and many will have a techno-tantrum as a result.

Therefore, we need to have something that will help bolster that level and calm their nervous system once they come off their device.

Green time is the best way to do this.

When kids are out in nature, it allows their brain to calm down, recalibrates their nervous system, gets them physically active and helps to overcome some of the potentially damaging effects of screen time. Many children these days are nature-starved. The local park, beach, lake, or even your backyard are obvious locations where you and your children can connect with nature. Even children living in major cities can play with sticks, roll in the grass, make mud pies and jump in puddles.

My sons love screens just as much as any of their friends and it can be a real struggle to manage their screen time. What has worked for us is to enrol the kids in some structured activities like a team sport, karate and even Scouts. One of my sons is a third generation Scout and while it can be a real challenge to pull him away from his gaming devices, he's actually happiest when he returns from a Scout camp dosed up on nature.

2. Educate your kids about sustainability

The second way you can nurture eco-conscious kids is to educate your children about sustainability.

In her book, *Reading Magic* Mem Fox states:

"If every parent understood the huge educational benefits and intense happiness brought about by reading aloud to their children, and if every parent—and every adult caring for a child—read aloud a minimum of three stories a day to the children in their lives, we could probably wipe out illiteracy within one generation."

Reading isn't just good to help develop literacy. Reading is great for education. Furthermore, reading books about the environment and sustainability to our children is one of the best ways to educate them in how to care for and protect their planet.

The growing number of books for children and teens focused on caring for the environment help nurture sustainability values within your child from an early age. I love the *Little Green Books* series for teaching younger children about sustainability and of course, Dr. Seuss' *The Lorax* is always a popular choice. Reading is also an activity that can be eco-friendly and free, just join your local community library to start enjoying the benefits.

3. Encourage eco-play

Given the global toy market reached almost USD95 billion in 2020[54], it's not surprising that many homes and landfills are overflowing with toys. Many commercial toys are designed to be short-lived and are easily broken and then discarded soon after purchase. Many of these are also made of plastic, which we know never breaks down in our environment.

Surely there are more sustainable ways to entertain our kids!

Moving from relying on commercial toys to eco-play activities does require creativity and planning but the benefits for your children, home and environment are worth the effort. With a little imagination and planning you can give kids a taste of some common eco-play activities from generations past and in the process gain a tidier lounge room floor and healthier family budget. A favourite eco-play activity of my sons when they were toddlers was emptying out my kitchen cupboards and having them match the lids to the containers. Turning cardboard boxes into cubby houses, trains, tunnels or boats provided hours and hours of entertainment. My sons also loved water painting. This simply involves giving your toddler a pail of water and a paint brush and sending them outside to paint anything they wish. It's the perfect toddler parent win-win, a fun craft activity with no mess to clean.

An eco-craft activity requiring minimal materials is collages. Most preschoolers love collages and you really only need glue and a paint brush to get started. Nature collages can be made from leaves, gumnuts, feathers and sand. If you've got some old catalogues, greeting cards or magazines lying around, these can be cut up to make colourful masterpieces and give new life to pre-used items in the process.

Libraries and garage sales can also be worth their weight in gold. I saved the annual membership fee for our local toy library many times over by borrowing various toys and puzzles. I especially loved the fact that as soon as my sons tired of a toy, I could replace it with another without adding clutter in my home. We also picked up a few toy bargains at local garage sales and our town's buy swap and sell social media site. And of course, passed on many of our kids' toys to charity stores and enjoyed knowing that we were giving the toy another lease of life to a family in need.

4. Lead by example

Put simply, how can we hope that our children grow up to live sustainably and care about the environment if we're not leading in this area? This is where we need to walk the talk and lead by example. It's not always an easy task but don't underestimate its effectiveness.

So tread lightly on the environment:

» Don't just not litter yourself, make an effort to pick up other people's litter when you're out and about.

» Make sure you're recycling and composting in your home and when you're out and about, if you can't find a recycling bin, just bring that recyclable item back home and place it in your kerbside recycling collection.

» Be a conscious consumer and be really mindful of what you purchase.

» Walk, bicycle or catch public transport wherever you can and make using your vehicle your last, not first, transport option.

» Set a technology example by managing your own screen usage and lead an active lifestyle yourself, or better still, be active together as a family.

» Create your sustainable home in the suburbs and explain to your children why you are making each change. This helps them learn along the way and become a part of family decisions and the functioning of your household.

5. Give kids a chance to shine

I totally understand that having your kids pitch in around the home initially takes a lot of patience, time and perhaps a few broken household items before they learn the ropes. But kids learn when you get them involved. Perhaps you can allocate your children a section in the garden that's their responsibility to grow what they want. Maybe they can collect the eggs from the chickens each day, or take out the rubbish or recycling, or be the light police and catch anyone out who leaves lights on in vacant rooms. If you have older children try and hand over the responsibility of cooking dinner one night a week to them. It's not cruel. It's not slave labour. It's smart and it's called insourcing.

I left home at the age of 17 to move to the city and study at university. I left knowing how to cook, clean and run a household. While I thought I was hard done by on many occasions growing up, I didn't fall flat on my face when I moved out of home. I just got on with living and cooked an awesome dinner from scratch and invited my university friends over to celebrate my newfound freedom!

Agreeing as a family that you will become more active together will reap so many benefits for your children and your family.

Final thoughts

Now I can understand that you might be thinking that this sounds great and it would be fantastic in an ideal world. But you're working, your partner's working and you literally don't have the time to entertain your kids, reduce their screen time, get them out in nature and still find time to exercise and cook a healthy dinner each night.

We're all stretched and time-poor, most families are dealing with screen-addicted kids, but this is a case of where the effort truly is trumped by the rewards. Starting small and setting screen time limits and agreeing as a family that you will become more active together will reap so many benefits for your children and your family as well:

- » Prioritise a screen-free Sunday and do a family activity like go on a picnic or explore your local environment.
- » Schedule some other activities in the times when your kids would otherwise be looking to be on their screens.
- » Perhaps enrol your kids in screen free activities like karate, a team sport, Scouts, dancing or learning a musical instrument.

All these activities will help build well-rounded kids, develop social and leadership skills in your children, as well as improve their physical and mental wellbeing.

The world has changed. Our streets have become busier, suburbs more congested and communities not as safe as they were when many of us were playing outside in our childhoods. Indeed, you're entitled to feel that there are fewer options for safe outdoor play. This is where I again recommend looking towards your local sporting clubs, Scouts or Guides. Of course, we don't want to over-schedule our kids as they need downtime and we also need downtime from running them around to extracurricular activities. But enrolling them in these activities can help share the load and you might find that the team coach or leader of the Scout troop becomes part of your extended village.

Like me, you may be raising your children away from where you grew up and you don't have grandparent support or uncles and aunties close on hand to help with parenting. It truly takes a village to raise a child and many of us just don't have that village support. But that doesn't mean that you can't build your own village. I used to attend a weekly playgroup run by one of my local churches. I looked forward to going there every week to be loved on by the grandparents who perhaps were separated by distance from their own families. They were only too happy to hold my babies and give me a break.

Don't feel guilty leaning on childcare, afterschool care, sharing the childcare load with other families or bringing in paid domestic help. By doing so, you can keep your career and family afloat and free up some time on a weekend for quality family time, rather than doing housework while your kids kick back on screens. Look for ways to build your village so you can create quality time to spend as a family.

Take action now!

It's time for you to take action to help nurture the eco-conscious kids in your life. Commit to completing the three actions below:

1. Firstly, set a screen time contract with your children and include a screen-free time for the entire family, such a screen-free Sundays, where you can spend some quality time together, preferably outdoors in nature.

2. Secondly, book a family outing in nature in your calendar each month for at least the next three months to help you get in the habit of spending this time together as a family. Give everyone in your household a chance to choose what activity they'd like to do. You can go on a picnic, a bike ride on a nature trail, take a short bushwalk or even try your hands at fishing. Just get out there and enjoy the great outdoors.

3. And lastly, if you haven't yet done so, join your local community library and borrow one sustainability-themed book for your child to read.

Parenting is a tough job. We literally are raising the next generation while we're working, managing a household and often caring for elderly family members as well. But by turning screen time into green time, educating your kids about sustainability, encouraging eco-play, leading by example and giving your kids a chance to shine, you are giving your children the best opportunity to improve their own physical and mental health. They will develop a love and respect for our environment and develop life and leadership skills which they'll draw upon in the future and for this, future generations will thank you all.

If you take action with regard to nurturing the eco-conscious kids in your life, take photos and tag me on social media @lauratrottadotcom. I'd love to see your progress!

CHAPTER 11
Invest your money ethically

> *"Where you invest and divest your money today creates the world of tomorrow."*
>
> ——Laura Trotta

Elizabeth Heyrick was one of the most prominent female campaigners against slavery in the 1820s. Disgusted over Britain's enslavement of people on the West Indies islands of Barbados and Jamaica (where large sugar plantations produced virtually all the sugar consumed in Western Europe), she took action. Heyrick set up a 'Ladies' Anti-Slavery Society' in Birmingham and launched a campaign to encourage Britons to quit using sugar produced in the West Indies and grocers not to stock the product. She suggested that if people wanted an alternate source of the "sweet dust", they purchase sugar grown in Britain's colonies in the East Indies—Bengal and Malaya—where cane field labourers were impoverished, but technically free.

Her campaign involved writing and distributing a series of educational pamphlets with phrases such as "Abstinence from one single article of luxury would annihilate the West Indian slavery!!"

While Heyrick died in 1830 before seeing her goal of "imminent emancipation" achieved, her efforts were not in vain. Samuel Sharpe, an enslaved Baptist deacon in Jamaica, read about the anti-slavery movement Heyrick did so much to fuel, and buoyed by the knowledge that many people in Britain were sympathetic to him and his peers, he formulated his own revolutionary vision. He then preached about it and his plans for rebellion to select groups of elite slaves.

Sharpe's rebellion, known as the Baptist War, ran for less than two weeks from 25 December 1831. It resulted in the destruction of dozens of buildings, the killing of hundreds of slaves and the hanging of Sharpe himself several months later. Despite the carnage, the demonstration of military competence made an impression like no other uprising had before and helped inspire the British Parliament to pass the Slavery Abolition Act 1833, which abolished slavery in the West Indies.

In short, news of the sugar boycott in headlines of 19th century newspapers crossed the Atlantic and helped inspire enslaved people to revolt. News of their visceral unhappiness to the point of mayhem helped inspire the British Parliament to push for immediate abolition—which is what Heyrick had been campaigning for all along.[55]

Boycotting those companies or products that don't align with your values can change the world. Likewise, for investing in ethical companies.

Whether it's opting for a green energy provider, eating organically grown food, avoiding fast fashion or divesting your financial investments from fossil fuels, tobacco or companies with a track record of modern slavery, conscious consumers and investors have long voted with their money and created positive change in the process.

In this chapter we'll take a look at how you can amplify your positive impact through ethical investing.

Why invest ethically?

Investing your money ethically has many benefits for your financial and personal wellbeing as well as our society.

Firstly, it allows you to have a clean conscience and to sleep soundly at night, knowing that you're not supporting industries that are causing harm to people and the planet. Your investments align with your values and the positive impact this has on your wellbeing cannot be denied.

Secondly, by investing in ethical and cleaner technologies you create a ripple effect, helping to reduce the cost of ethical products and cleaner technologies for others to follow.

Thirdly, developing cleaner technologies and purchasing products manufactured in safe factories by adults paid a fair

A majority of sustainable funds have outperformed traditional funds over multiple time horizons.

wage should be the norm, not the exception. By buying and investing in ethical products and businesses you are helping to make ethical business processes mainstream.

And finally, ethical investing is good for your wallet! Put simply, a detailed study of environmentally sustainable funds by Morningstar in 2020 revealed that they have outperformed traditional funds across the board—beating them during the COVID-19 pandemic as well as during the 10 years up to and including the coronavirus sell-off. Morningstar examined 745 sustainable funds and compared them against 4,150 traditional funds, and found they matched or beat returns in all categories, whether bonds or shares, UK or abroad. The study suggested that there is no performance trade-off associated with sustainable funds. In fact, a majority of sustainable funds have outperformed traditional funds over multiple time horizons.[56] A separate study by MSCI (Morgan Stanley Capital International) found portfolios that integrated ESG (environmental, social and governance) factors as well as financial analysis actually had lower risks and outperformed over time.[57]

What is ethical investing?

So you now understand why ethical investing is good for you and our society, but what does it mean exactly?

Ethical investing refers to investors using their money to drive social and environmental change. Ethical investing means different things to different people, but as a basic rule it is aligning personal values with your money. For example, some investors might be keen on renewable energy while others might want to simply exclude investments making a negative impact, like the tobacco industry. Investors who have an ethical focus will invest based on ESG criteria, choosing investments that align with their values and excluding those that do not align based on these metrics.

According to Investopedia, **environmental** criteria may include climate policies, energy use, waste, pollution, natural resource conservation and treatment of animals. The criteria can also help evaluate any environmental risks a company might face and how the company is managing those risks. Considerations may include direct and indirect greenhouse gas emissions, management of toxic waste and compliance with environmental regulations.

Social criteria focus on the company's relationships with stakeholders including suppliers, charitable organisations, its community and its workforce. Criteria may include if the company operates an ethical supply chain, supports LGBTIQ+ rights, has policies to protect against sexual misconduct and pays fair wages.

Governance ensures a company uses accurate and transparent accounting methods, pursues integrity and diversity in selecting its leadership and is accountable to shareholders. ESG investors may require assurances that companies avoid conflicts of interest in their choice of board members and senior executives, don't use political contributions to obtain preferential treatment, or engage in illegal conduct.[58]

When global resources company BHP's chairman Ken MacKenzie tried to quantify the rising influence of ESG, he painted a picture of how the seating arrangements have changed over the past decade at the regular meetings between companies and institutional investors.

A decade ago, the ESG analyst sat at the back of the room, quietly listening and taking notes on the conversation between the chief investment officer and the chief executive.

Within five years, the ESG analyst had earned a seat at the table and the right to ask a couple of questions.

Nowadays, MacKenzie says the ESG analysts hold court as the stars of the show.[59]

Big business is taking ESG seriously because they know they need to in order to survive. What led to them taking it seriously? Changing community, social and environmental values, evident in the behaviour of investors like you who sent them a very clear message through where you choose to, and just as importantly, choose not to, invest your money.

How to invest ethically

There is no rulebook to follow when investing your money ethically, but here are five steps to help get you started.

But first, please note that I am not a financial planner. Any investment thoughts or advice below are based on my opinion and general experience, not your specific case. As such I recommend you always seek independent financial advice

I wanted to ensure my superannuation was not invested in fossil fuels and instead, ensure that it's invested in renewable energy technologies.

for your circumstances before acting on any information presented.

1. Work out what you want to achieve through your investments and divestments

When getting started in ethical investing you'll first need to work out what you want to invest in and/or divest from. Christopher Zinn, finance specialist and consumer campaigner at Life Sherpa recommends:

> "The first thing, when you are looking at ethical investing, is to not just look at the financial products, which say they're ethical, but actually ask yourself what do you want to achieve, what are you interested in? It might be climate change. It might be modern day slavery. It might be something else very specific."

2. Switch your superannuation to an ESG fund

As my knowledge about climate change has grown, so too have the actions I've taken to not only decarbonise my own home, but to decarbonise my investments. Since I don't own many shares outside of superannuation, it made sense for me to focus on how my superannuation is invested. Specifically, I wanted to ensure my superannuation was not invested in fossil fuels and instead, ensure that it's invested in renewable energy technologies.

I researched different superannuation funds, drawing on Australia's Money Magazine Best of the Best recommendations for ESG superannuation funds, until I found one that fitted with my values. I then made the switch from a balanced fund with a reputable superannuation provider I'd been with for almost 20 years to a high-performing environmental investments fund with an equally reputable superannuation provider. Once I'd decided to make the switch, the process literally took me less than an hour and without leaving home. The process was much quicker and simpler than purchasing a new mobile phone and moving my contacts across!

Christina Hobbs, CEO and co-founder of Verve Super offers the following three pieces of advice when switching to an ethical superannuation fund:

> "Firstly, make sorting out your super a priority this weekend. It's actually not that big a task to do. Once you've made a decision, it could only take two minutes to make the switch.
>
> Secondly, if you are interested in making an ethical decision, it can be a really interesting exercise to do as a family. Sit down with your family and ask questions like, what do we really care about and what are the dealbreakers for us? It can lead to some very interesting discussions about ethics and values.
>
> Thirdly, look at the performance of the super fund—**not just fees**. There is so much focus on fees in the media, but not on performance. Learn to read and compare the performances between funds so you can pick one that is not only ethical but also performing well."

3. Purchase shares in ethical companies

The most common way to invest in ethical companies is to purchase shares of that company. To do this, find a share trading platform that lets you filter for ESG stocks and review how well the shortlisted companies are performing in relation to ESG prior to making your investing decision.

You can screen an ESG investment in two ways: negative screening or positive screening.

Negative screening simply involves excluding companies that fail ESG metrics. What you personally choose to negatively screen is up to you. Most commonly, negative screening will exclude companies with poor environmental performance or products that do societal harm such as gambling, guns or tobacco.

On the other hand, **positive screening** involves choosing investments with high ESG scores. Investors might focus on some or all aspects involved in ESG such as clean energy sustainable products, strong inclusion processes for employees and energy efficiency.

In order to gain an accurate measure of a company's ESG performance you can review company reports and/or third-party sources.

Companies are required to release reports detailing their performance twice per year. These reports allow investors to gain a snapshot of those businesses' operations. Businesses also host an annual general meeting (AGM) which provides

There are a number of key resources that can help investors choose more ethical companies.

the opportunity for investors to hear from the board of directors as well as engage with them.

It is not mandatory in Australia for businesses to comment on their ESG ratings when they are releasing reports or during an AGM, but most businesses that are doing the right thing don't shy away from reporting their ESG performance.

Secondly, there are a number of key resources that can help investors choose more ethical companies. In Australia, organisations such as the Responsible Investment Association Australasia (RIAA) can help investors make more informed ESG decisions. For investors outside Australia, MSCI ESG Ratings and Sustainalytics ESG Risk Ratings can help investors make informed decisions.

4. Purchase enough shares in unethical companies to buy a seat at the decision-making table

Another way to create significant change through investing is to buy enough shares in a company you wish to see change to give you a seat at the decision-making table for that company. Now this isn't a realistic option for most of us, but it's working for billionaire climate activists.

In 2022, the board of energy giant AGL (Australia's biggest carbon emitter) rejected an $8 billion bid from Mike Cannon-Brookes (Atlassian co-founder and climate activist) and Canadian giant Brookfield to buy the company and accelerate its exit from coal. The board intended to proceed with plans to demerge AGL.

Cannon-Brookes was against the demerger as he believed it would destroy shareholder value and leave the company as two smaller entities, less able to bankroll the big investments needed to bring forward the closures of its coal-fired power stations that are currently not due to retire until 2045.

To prevent the demerger, Cannon-Brookes' private investment company Grok Ventures quickly amassed an 11.3% interest in AGL and used that stake to convince other investors to vote against the demerger. Within three days, the AGL demerger was dead in the water.

The fallout included the resignation of the managing director, chairman and two independent directors. Two remaining directors were set to conduct a strategic review into the future of the 180-year-old company.

Cannon-Brookes' response was simple: "Huge day for Australia ... had to sit down and take it in."

Ethical investors across the entire country also had to sit down and take it in. Cannon-Brookes, through his bold investing had rolled AGL's demerger, just a month after his failed bid to purchase the company outright.

Who knows which company he'll crash tackle next for the benefit of our planet?

5. Let your everyday banking create the future you want!

If you don't have any idle money to invest you can still make a positive impact through your everyday banking activities. I know my football-mad son is much more inclined to save his hard-earned pocket money knowing that his bank is contributing financially to his local football club!

Many community-owned banks are certified B Corporations. B Corporations are required to focus on social and environmental performance as well as transparency and accountability. B Corporations aim to be a force for good by looking at the impact of their business on broader stakeholders including customers, community and environment. All B Corporations are individually assessed every three years by B Lab, a not-for-profit organisation that manages the certification.

It's normal to have some reservations when changing your investments, superannuation funds and even everyday banking accounts to more ethical alternatives. The most common is concern about the investment return not being as high. Christopher Zinn, advises not to switch to an ethical fund or finance product to chase a greater return, because that may or may not happen. It may or may not happen with any fund. One of the issues around ethical funds is diversification. With a standard fund, there are a whole lot of industries that you might be engaged with like resources, but ethical stocks can tend to be more heavily weighted to

*Don't underestimate
the power you have
as an investor.*

technology. Now that can be good because technology's been very buoyant, but if there's a bust and you have all your eggs in one basket that can be painful. Costs can also be higher because of high management fees due to ethical funds doing extra screening work.

You may also be concerned about the time required to research your ethical investment options and the hassle involved in making the switch. Once I decided which superannuation fund to switch to the process literally took less than an hour to set up a new account, complete authorisation to transfer my superannuation from my old fund into the new fund and complete notification to my employer to change my superannuation fund. Most banks have automated processes these days to facilitate a seamless set-up of accounts with recurring payments.

And lastly, if you're worried about greenwashing or falling for scams, I simply say be guided by the experts and let them do the work for you! Either seek the advice of your personal financial advisor and/or consult resources such as Australia's *Money Magazine's* annual Best of the Best recommendations for ESG managed funds, bank accounts and superannuation funds.

Final thoughts

Don't underestimate the power you have as an investor. Money speaks volumes in our modern world. By divesting away from companies and industries you don't wish to support and investing in those that create a better future, you're not only sending a strong message to the market, you're amplifying your positive impact beyond your suburb.

Take action now!

It's time to put your money where your mouth is and take action on ethical investing. Get started today by taking these simple steps:

1. Decide what you want to invest in or divest from i.e., what is ethical or unethical to you?

2. Ensure your superannuation is invested ethically and if it isn't, investigate making the switch from your current fund into an ESG fund or ESG product within your current superannuation fund. If you're investing in shares outside superannuation, use a trading platform and filter for ESG stocks.

3. Ensure your everyday banking is ethical and is improving outcomes in your local community rather than lining the pockets of executives and shareholders. Switch to a community-owned B Corporation bank if you haven't already done so!

Share what action you've taken to invest for the future on social media with the hashtag #sustainabilityinthesuburbs and tag me @lauratrottadotcom so I can personally congratulate and thank you.

CHAPTER 12
Demand a better future

"Vote like your life depends on it, because it does."
—AL GORE

When I turned 18 and was legally permitted to vote in Australian elections, I saw it as an inconvenience. I had no interest in politics and tuned out whenever arguing politicians were featured in the nightly news bulletin. It was no surprise then that when election day came around, I had no idea who to give my vote. Ironically, my best friend was very interested in politics and couldn't understand my indifference. He was studying economics and would host election parties with his university friends where they'd watch the live election coverage on television and cheer when the party they were going for would win a seat. This seemed strange to me; I saw elections (and election parties) as something to endure, not enjoy.

Around this time, I also believed that working hard in a science and engineering field within heavy industry to improve their environmental performance was the most effective way for me to enact change. It was what led me to working within the resources sector. I've learnt that while this does generate change, it can be painstakingly slow and extremely frustrating at times.

Years have passed since I reluctantly placed my first vote and I've grown older and wiser. I now see stable government and responsible economic management as being critical for strong environmental management. I also believe that engaging with politics is one of the most effective ways to enact environmental change on a grand scale—no hard hat or steel capped boots required! Furthermore, I now follow politics closely and view voting as the immense privilege it is.

It would appear I'm not the only one who's changed. Almost three decades ago when I was enduring those election parties, most big business and CEOs of big business were primarily concerned about the government's economic management. They just wanted to deliver a strong return on investment to their shareholders. Environmental issues were rarely mentioned. By comparison, a survey conducted in early 2022 of Australia's C-suite found that economic management and climate change topped the list of issues most concerning to company directors for the 2022 federal election, with company directors unhappy about a lack of long-term policy vision for both.[60] It's fair to say that today's company directors are just as concerned about environmental management as they are economic management. The two are no longer mutually exclusive.

Just as we all have power to improve the health of our environment through our individual actions, those of us in democratic societies have the power to make positive environmental change through our individual vote and how we demand action from our leaders on environmental issues and the climate crisis.

Change happens when enough people demand it.

Demanding what you want isn't always simple. It involves first deciding what you want and then finding the courage to ask for it. But when we know what we want, confidently ask for it and get it, we have an immense feeling of pride that we matter and what we want is important.

You don't need me to remind you of the importance of climate action, preserving our natural environment, halting biodiversity loss and reducing pollution.

In this chapter, I share three ways that you can use your voice to actively demand a better world for current and future generations.

1. Vote for the future

Firstly, we need to vote for the future, and this involves letting candidates in power know what you want and then voting for what you want.

Research your candidates and the environmental and climate policies of their political parties to help inform your vote.

I live in the seat of Boothby, one of the most marginal federal seats in Australia. It was held by the Liberal Party with a 1.4% margin heading into the 2022 election and swung to the Labor Party in 2022 for the first time since the Second World War.

In the lead up to the 2022 Australian federal election, I attended several online forums where my local candidates presented their policies and answered questions from the audience. I attended these forums to ask candidates what action they would take on issues such as climate change. This helped me form an opinion about which candidate best aligned with the issues of most concern to me. Attending these forums also helps the potential candidates know what issues their communities are most concerned about so they can best represent them in parliament.

Research your candidates and the environmental and climate policies of their political parties to help inform your vote and then follow through with your vote for action on climate and the environment. Australians did this en masse in the 2022 federal election, voting for climate action like they'd never done before and sending the Coalition government into opposition. The election result also clearly told our former prime minister Scott Morrison, who only five years earlier brought the lump of coal into parliament, that we are indeed very scared of the impact of continually extracting and burning coal and we're more than ready to decarbonise our society.

My local representative regularly sends out surveys seeking feedback on the most important issues concerning voters in the electorate. I make a concerted effort to always complete these surveys to let the representative know that I'm concerned about climate change and that I expect him or her to lead on climate action. In the absence of surveys, you can write to your local member of parliament to let them know that strong environmental management and action on climate change is important to you and that you will be voting for the candidates who can deliver the best policies in this area.

If you are too young to vote, you can still get involved.

Lola Stravoskoufis, a Year Nine student from Exeter in the Southern Highlands region of New South Wales, marched as a Year Six student in the large student climate strikes in 2019 and subsequently joined the School Strike 4 Climate movement. She now leads the School Strike 4 Climate chapter in her town and actively met with her local candidates for the federal election to ensure that the voice of youth was heard loud and clear. So, while she legally couldn't vote, she still played an active role and ensured that her local state and federal representatives knew and understood what issues were most important to the youth, voters in the not-too-distant future.

2. March for the future

In 2018, after experiencing Sweden's hottest summer on record, a 15-year-old Greta Thunberg protested alone, outside the Swedish parliament with a sign saying

"Skolstrejk for Klimatet" (School strike for climate) to pressure her government to meet carbon emission reduction targets. Within a few short months, millions of students followed her lead, taking Fridays off school to protest for climate action.

History has shown that when approximately 3.5% of the population participates in non-violent protests, change becomes inevitable. For Australia's population of 26 million, that's less than 1 million people marching, which the 2019 climate strikes weren't too far off reaching. Those strikes alone attracted more than 300,000 protesters in Australian cities, and I'm sure if it weren't for lockdowns associated with the COVID-19 pandemic, the numbers of protesters marching for climate action in 2020 and 2021 would have been several times that again. There's been a long history of protests enacting great change in environmental movements

Observing a Jabiluka mine protest in 1998

History has shown that when approximately 3.5% of the population participates in non-violent protests, change becomes inevitable.

around the world, including Australia. Possibly the most noteworthy was the Franklin Dam project proposed on the Gordon River in Tasmania in the 1980s. The dam site was occupied by protesters in December 1982, leading to widespread arrests and extensive publicity. The issue featured in the 1983 federal election, helping to bring down the then-government of Malcolm Fraser. The new government, led by Bob Hawke, had promised to stop the dam from being built. A legal battle between the federal and Tasmanian governments followed, resulting in a landmark High Court ruling in favour of the federal government and "No Dams" advocates. Many Franklin Dam campaigners, notably Bob Brown and Christine Milne, went on to form the Australian Greens political party in 1992.

My first environmental protest was a march in Melbourne against the development of the Jabiluka uranium mine adjacent to the world listed Kakadu National Park in the Northern Territory, a significant environmental issue in the late 1990s. I chose to study the issue of the mine development for an assignment in an environmental ethics subject at university and attended the protest to learn more about the issue.

Jabiluka is a pair of uranium deposits in the Northern Territory of Australia. The mine was to have been built by Energy Resources Australia (ERA) on land belonging to the Mirarr Traditional Owners. In 1998, activists came from across Australia and abroad to blockade the construction of the mine. Over 500 people were arrested during the eight-month blockade. Despite the protest and the blockade, ERA proceeded with the development of the mine, erecting surface infrastructure and a decline down to the ore body to allow for further delineation of the resource. When uranium prices started to fall, the project was stalled. In 2000, ERA's parent company North Limited was taken over by Rio Tinto group, who announced that the mine would not proceed.

The Mirarr Traditional Owners continually requested that Rio Tinto clean up the mine site, and in 2003, 50,000 tonnes of excavated material was put back down the Jabiluka decline, works to remove and remediate the interim border management pond commenced, and over 16,000 native plants were planted in rehabilitation efforts. Even though future economic drivers may still lead to increasing pressure for the extraction of the uranium that has remained in the ground, Jabiluka is another example of where advocacy for environmental and indigenous rights resulted in a hard-fought win.

Maybe you've considered marching or protesting about climate action or a local environmental issue, but you're worried about getting arrested. Or perhaps, being a 'nuisance' just isn't in your DNA. I'm not suggesting or encouraging you to glue yourself to the footpath outside the office of your local oil and gas energy giant. Rather, choose a method of protest you're comfortable with to demand action on. I'm personally not one for chaining myself to old growth trees that are to be felled or coal trains headed for export, and I'm certainly not telling you to do so. (However, I can appreciate how people can feel so desperate for environmental or climate action that they believe they've got no other choice but to put their body on the line and risk getting a criminal record.)

There comes a point where governments can't ignore thousands of people marching for change. Thousands of peaceful protesters filling our city streets send a really strong message to leaders and decision makers about what people want. I do acknowledge that in some areas of the world participating in peaceful marches or undertaking any acts of disobedience is not possible or even safe. This only further highlights the importance for those privileged to live in a society where free speech is tolerated, to maximise this right and march for environmental and climate action.

3. Divest and invest for the future

As discussed extensively in the previous chapter, the third way you can demand a better future is to divest (e.g. from fossil fuels) and invest for the future (e.g. in renewables). Divest from companies that are also engaging in lobbying against citizen action on climate change and vote with your money. Stop buying their stocks, stop buying their products and services where alternatives exist. Look into what banks you use, what insurance products you buy, where your superannuation is invested and make the change where you find that they are invested in old, dirty technologies. If you can, and if you're financially able to, switch your investments to future renewable technologies and technology that's going to drive a cleaner future.

Respect democracy and respect the laws of your society in exerting your freedom of thought.

Final thoughts

We need strong government and institutions in place to deal with the environmental and climate challenges ahead. Governments that are stable and are trying to find ways to meet these challenges should be worked with, not dismantled. Those of us with the privilege of being able to vote have a responsibility to leverage our influence as far as we can, but respect democracy and respect the laws of your society in exerting your freedom of thought. So take care and stay safe and responsible while you demand and vote for a better future.

Take action now!

It's over to you to demand more and to vote for the future you want to see. You can start by taking the following two actions:

1. Let your local member of parliament know by commenting on their social media posts and/ or writing them a letter or email that strong environmental management and action on climate change is important to you.

2. Secondly, get active and support peaceful environmental rallies in your community. Perhaps you could follow School Strike 4 Climate and participate in the next climate march in your area. Don't leave it up to the kids to demand action on climate from world leaders. March as a responsible adult and demand that adults

take the urgent action required to improve environmental management and address the climate crisis.

Share what action you've taken to demand a better future on social media with hashtag #sustainabilityinthesuburbs and tag me @lauratrottadotcom so I can personally congratulate and thank you.

CONCLUSION

We've covered a lot together in this book.

I hope you now feel better equipped to take action to reduce your environmental footprint and live more sustainably.

Let's recap what we've covered ...

First, I shared how we got here and why we need to change, including the top seven environmental issues facing our planet today. Were you surprised that I placed overpopulation at the top of the list?

In Part 1 I shared some tips to help you manage eco-anxiety and move through climate grief. Sustainable self-care was front and centre because more than anything I believe you need to bed down those habits before you launch into saving the world, otherwise you'll be on the express train to burnout with no energy for yourself let alone our precious planet.

In Part 2 we unpacked how to reduce your impact by considering ecotarianism, growing your own food and creating a zero-waste home and lifestyle. Minimalism, toxin-free living, decarbonising your home and becoming water smart then followed to complete this practical section.

In Part 3 you learnt three ways to expand your impact by raising eco-conscious kids (or positively influence the children in your life if you have none of your own!), investing for the future and demanding a better future.

We've covered a lot, but we're certainly not done!

There are so many of us on this planet and in future years there will be many more of us. We have no choice but to change the way we are living on and treating Earth. And there's much that each of us can do.

I encourage you to leave this book in an easy to reach place so you can continue to flip through it in coming months and years. I'm sure you'll find one small thing you can change each time you dive into its pages.

Imagine if everyone did one small thing to improve our environment today. What a positive impact that would make!

So, go.

Choose your one small thing and go do that thing right now.

Then choose another small thing tomorrow or next week.

Then choose another, and another.

Before too long your momentum will carry you into your new sustainable life and the changes you're making will not only become habit, you'll create a ripple effect that will be felt beyond your suburb, city and country.

On behalf of future generations of all lifeforms on Earth, I thank you for starting with your one small thing today.

Together we can make green mainstream.

ADDITIONAL INFORMATION AND RESOURCES

Climate & Mind: **climateandmind.org/where-to-find-help**

The Climate Reality Project: **climaterealityproject.org**

The Environmental Working Group: **ewg.org**

School Strike 4 Climate: **schoolstrike4climate.com**

B Corporation: **bcorporation.net**

Verve Super: **vervesuper.com.au**

Money Magazine Australia: **moneymag.com.au**

Responsible Investment Association Australasia (RIAA): **responsibleinvestment.org**

MSCI ESG Ratings: **msci.com/our-solutions/esg-investing/esg-ratings**

Sustainalytics: **sustainalytics.com**

ENDNOTES

1   NASA 2022a, Data for December 2021, available online at *https://gml. noaa.gov/ccgg/trends/* [accessed 11 May 2022]

2   NASA 2022b, Global Climate Change: Vital Signs of the Planet, Global Temperature, *https://climate.nasa.gov/vital-signs/global-temperature/* [accessed 20 August 2022]

3   NASA Earth Observatory, *https://earthobservatory.nasa.gov/world-of-change/global-temperatures*, [accessed 10 July 2022]

4   Reserve Bank of Australia, 19 September 2019, The Changing Global Market for Australian Coal, *https://www.rba.gov.au/publications/ bulletin/2019/sep/the-changing-global-market-for-australian-coal.html* [accessed 17 August 2022]

5   Parliament of Australia, Retirement of Coal Fired Power Stations Interim Report Chapter 2: Electricity markets and the role of coal fire power stations, *https://www.aph.gov.au/Parliamentary_Business/ Committees/Senate/Environment_and_Communications/Coal_fired_power_ stations/Interim%20Report/c02* [accessed 17 August 2022]

6   Nordhaus, T, 5 July 2018, The Earth's carrying capacity for human life is not fixed, *https://aeon.co/ideas/the-earths-carrying-capacity-for-human-life-is-not-fixed* [accessed 17 August 2022]

7   United Nations, Department of Economic and Social Affairs, 26 February 2022, UN DESA Policy Brief No. 130: Who population growth matters for sustainable development, *https://www.un.org/development/desa/dpad/publication/un-desa-policy-brief-no-130-why-population-growth-matters-for-sustainable-development/* [accessed 10 July 2022]

8   NASA Earth Observatory, *https://earthobservatory.nasa.gov/world-of-change/global-temperatures*, [accessed 10 July 2022]

9   CSIRO and BOM 2020, State of the Climate 2020, at *https://www.csiro.au/en/research/environmental-impacts/climate-change/state-of-the-climate* [accessed 10 July 2022]

10   United Nations, 19 January 2022, UN News, 2021 joins top 7 warmest years on record: WMO, *https://news.un.org/en/story/2022/01/1110022* [accessed 10 July 2022]

11   IUCN Red List, Background & History, *https://www.iucnredlist.org/about/background-history* [accessed 17 August 2022]

12   Mongabay, 22 July 2012, Saving what remains – Medicinal Plants, *https://rainforests.mongabay.com/1007.htm* [accessed 17 August 2022]

13   UNICEF, Water, *https://www.unicef.org/wash/water* [accessed 17 August 2022]

14   European Environment Agency, Ocean Acidification, *https://www.eea.europa.eu/ims/ocean-acidification* [accessed 17 August 2022]

15   Cheapa Waste Skips, Global Waste Statistics 2022, *https://cheapawasteskips.com.au/global-waste-statistics-2022/* [accessed 17 August 2022]

16   The Numbers, An Inconvenient Truth (2006), *https://www. the-numbers.com/movie/Inconvenient-Truth-An* [accessed 13 May 2022]

17   American Psychological Association, Majority of US Adults Believe Climate Change is Most Important Issue Today, 6 February 2020, *https://www.apa.org/news/press/releases/2020/02/ climate-change*, accessed 19 April 2022.

18   Climate & Mind, What is Climate Grief? *https://www. climateandmind.org/what-is-climate-grief* [accessed 21 April 2022]

19   PSYCOM, Climate Grief: The Emotional Toll of Climate Change, *https://www.psycom.net/anxiety/coping-climate-grief-anxiety* [accessed 19 April 2022]

20   Cunsolo, A., & Ellis, N. R. (2018). Ecological grief as a mental health response to climate change-related loss. Nature Climate Change, 8(4), 275.

21   The Conversation, Against the odds, South Australia is a renewable energy powerhouse. How on Earth did they do it?, *https://theconversation.com/against-the-odds-south-australia-is-a-renewable-energy-powerhouse-how-on-earth-did-they-do-it-153789* [accessed 21 April 2022]

22   Hornsdale Power Reserve, South Australia's Big Battery, *https:// hornsdalepowerreserve.com.au/* [accessed 19 April 2022]

23   Renew Economy, South Australia looks to more big batteries to maximise wind and solar exports, *https://reneweconomy.com. au/south-australia-looks-to-more-big-batteries-to-maximise-wind-and-solar-exports/* [accessed 21 April 2022]

24   Men's Health, The '20-5-3' Rule Prescribes How Much Time to Spend Outside, *https://www.menshealth.com/fitness/a36547849/ how-much-time-should-i-spend-outside/* [accessed 22 April 2022]

25   Henry Ford Health, 14 December 2021, Screen Time Limits Aren't Just For Kids. Why Adults Need Them Too, https://www. henryford.com/blog/2021/12/adult-screen-time-limits [accessed 27 April 2022]

26   Harvard Health Publishing, 7 July 2020, Blue Light Has a Dark Side, https://www.health.harvard.edu/staying-healthy/blue-light-has-a-dark-side [accessed 27 April 2022

27   Mayo Clinic, Exercise and stress: Get moving to manage stress, https://www.mayoclinic.org/healthy-lifestyle/stress-management/in-depth/exercise-and-stress/art-20044469 [accessed 22 April 2022]

28   Sleep Foundation, Why Do We Need Sleep?, https://www. sleepfoundation.org/how-sleep-works/why-do-we-need-sleep [accessed 27 April 2022]

29   Oxford Academic European Heart Journal, 9 November 2021, Accelerometer-derived sleep onset timing and cardiovascular disease incidence: a UK Biobank cohort study, https://academic. oup.com/ehjdh/article/2/4/658/6423198 [accessed 27 April 2022]

30   Kettering Global, 15 April 2019, Why Hobbies Are Important, https://online.kettering.edu/news/2019/04/15/why-hobbies-are-important [accessed 27 April 2022

31   Cunha LF, Pellanda LC, Reppold CT. Positive psychology and gratitude interventions: a randomized clinical trial. Front Psychol. 2019;10:584. doi:10.3389/fpsyg.2019.00584

32   Department of Primary Industries and Regional Development, 11 October 2021, Reducing livestock greenhouse gas emissions, https://www.agric.wa.gov.au/climate-change/reducing-livestock-greenhouse-gas-emissions [accessed 19 June 2022]

33   Kopp, W, 24 October 2019, National Library of Medicine, How Western Diet and Lifestyle Drive the Pandemic of Obesity and

Civilization Diseases, https://www.ncbi.nlm.nih.gov/pmc/articles/ PMC6817492/ [accessed 12 September 2022]

34   Crinnion, W, 15 April 2010, National Library of Medicine, Organic foods contain higher levels of certain nutrients, lower levels of pesticides, and may provide health benefits for the consumer, https://pubmed.ncbi.nlm.nih.gov/20359265/ [accessed 19 June 2022]

35   Environmental Working Group, 7 April 2022, EWG's 2022 Shopper's Guide to Pesticides in Produce, https://www.ewg.org/ foodnews/summary.php [accessed 19 June 2022]

36   Choice, 19 Sep 2014, Fresh food tricks, https://www.choice.com.au/ shopping/everyday-shopping/supermarkets/articles/fresh-food-tricks [accessed 19 June 2022]

37   Eartheasy, 19 June 2021, Six Unexpected Health Benefits of Gardening, http://learn.eartheasy.com/2014/09/6-unexpected-health-benefits-of-gardening/ [accessed 28 August 2016]

38   Clean Up Australia, Australia's waste challenges go far beyond one day, https://www.cleanup.org.au/clean-up-our-waste [accessed 22 June 2022]

39   Clean Up Australia, Australia's waste challenges go far beyond one day, https://www.cleanup.org.au/clean-up-our-waste [accessed 22 June 2022]

40   Panthenon Enterprises, The Story Behind "Reduce, Reuse, Recycle", http://pantheonchemical.com/reduce-reuse-recycle/ [accessed 22 June 2022]

41   Kilvert, N, 14 October 2021, E-waste surges in 2021 as world sends goldmine to landfill, https://www.abc.net.au/news/ science/2021-10-14/e-waste-electronics-landfill-gold-landfill-recycling/100524744 [accessed 1 July 2022

42   Foodbank, Food Waste Facts, https://www.foodbank.org.au/food-waste-facts-in-australia/ [accessed 21 August 2022]

43   Mohan, Mirela, 15 March 2022, More than a third of Americans rent self storage, with furniture the most stored item, https://www.storagecafe.com/blog/self-storage-use-and-main-demand-drivers/ [accessed 2 July 2022]

44   Self Storage Australia, 15 February 2022, Four interesting facts about the self storage industry, https://www.selfstore.com.au/4-interesting-facts-about-the-self-storage-industry/ [accessed 2 July 2022]

45   Jeffrey J. Froh, Christopher J. Fives, J. Ryan Fuller, Matthew D. Jacofsky, Mark D. Terjesen & Charles Yurkewicz (2007) Interpersonal relationships and irrationality as predictors of life satisfaction, The Journal of Positive Psychology, 2:1, 29-39, DOI: 10.1080/17439760601069051 [accessed 2 July 2022]

46   Choi H, Schmidbauer N, Sundell J, Hasselgren M, Spengler J, Bornehag C-G (2010) Common Household Chemicals and the Allergy Risks in Pre-School Age Children. PLoS ONE 5(10): e13423. https://doi.org/10.1371/journal.pone.0013423 [accessed 2 July 2022]

47   Phys Org, 24 January 2019, South Australia heatwave smashes record temperatures, https://phys.org/news/2019-01-south-australia-heatwave-temperatures.html [accessed 12 September 2022]

48   Unicef, Water scarcity – Addressing the growing lack of available water to meet children's needs, https://www.unicef.org/wash/water-scarcity [accessed 9 July 2022]

49   Home water works, Toilet Water Saving Tips, http://www.home-water-works.org/indoor-use/toilets [accessed 6 July 2022]

50 Australian Bureau of Statistics, 20 October 2021, Water Account, Australia, https://www.abs.gov.au/statistics/environment/ environmental-management/water-account-australia/latest-release [accessed 9 July 2022]

51 Ribner, AD, Coulanges, L, Friedman, S, Libertus, ME, National Library of Medicine, Screen Time in the Coronavirus 2019 Era: International Trends of Increasing Use Among 3- to 7- Year Old Children, https://pubmed.ncbi.nlm.nih.gov/34461061/ [accessed 25 May 2022]

52 Paulich KN, Ross JM, Lessem JM, Hewitt JK (2021) Screen time and early adolescent mental health, academic, and social outcomes in 9- and 10- year old children: Utilizing the Adolescent Brain Cognitive Development (ABCD) Study. PLoS ONE 16(9): e0256591. https://doi.org/10.1371/journal.pone.0256591 [accessed 25 May 2022]

53 Australian Government Department of Health, 2021, Physical activity and exercise guidelines for all Australians, https://www. health.gov.au/health-topics/physical-activity-and-exercise/physical-activity-and-exercise-guidelines-for-all-australians [accessed 25 May 2022]

54 Statistica, Global toy market value increases for the seventh year running, https://www.statista.com/markets/415/topic/1001/ toys/#statistic1 [accessed 25 May 2022]

55 Zoellner, T, 10 July 2020, The Conversation, How one woman pulled off the first consumer boycott – and helped inspire the British to abolish slavery, https://theconversation.com/how-one-woman-pulled-off-the-first-consumer-boycott-and-helped-inspire-the-british-to-abolish-slavery-140313 [accessed 18 June 2022]

56   Collinson, P, 13 June 2020, The Guardian, Ethical investments are outperforming traditional funds, https://www.theguardian.com/money/2020/jun/13/ethical-investments-are-outperforming-traditional-funds [accessed 18 June 2022]

57   Micallef, C, 14 January 2022, Finder, How to start investing ethically, https://www.finder.com.au/ethical-investing#pros-and-cons-of-esg-investing [accessed 18 June 2022]

58   Investopedia, 18 May 2022, Environmental, Social, and Governance (ESG) Criteria, https://www.investopedia.com/terms/e/environmental-social-and-governance-esg-criteria.asp [accessed 18 June 2022]

59   Ker, P, 16 March 2022, Australian Financial Review, Can the 'Wild West' of ethical investment be tamed?' https://www.afr.com/policy/energy-and-climate/can-the-wild-west-of-ethical-investment-be-tamed-20220310-p5a3jw [accessed 18 June 2022]

60   Climate, economy top list of C-suite election issues, 20 April 2022, https://www.afr.com/politics/federal/climate-economy-top-list-of-c-suite-election-issues-20220420-p5aetr [accessed 18 May 2022

ABOUT THE AUTHOR

Laura Trotta, MSc, BEng (Environmental) (1st class Hons), is an experienced environmental engineer and award-winning sustainability educator.

Laura is a passionate believer in addressing the small things to achieve big change and protecting the planet in practical ways. She has inspired and educated tens of thousands of people globally to adopt a greener lifestyle through her popular 'Eco Chat' podcast and award-winning sustainable living programs Home Detox Boot Camp and Sustainability in the Suburbs.

Laura lives in Adelaide, South Australia with her husband, Paul and their two children, Matthew and Christopher.

lauratrotta.com
facebook.com/lauratrottadotcom
linkedin.com/in/lauratrottadotcom
instagram.com/lauratrottadotcom